HALLEY'S COMET,
1755-1984

Rendition of Halley's Comet (1456) as it appeared in the *Nuremberg Chronicle*. This illustration is commonly seen in the literature of Halley's Comet.

Halley's Comet, 1755-1984

A BIBLIOGRAPHY

Compiled by
Bruce Morton

GREENWOOD PRESS
Westport, Connecticut • London, England

Library of Congress Cataloging in Publication Data

Morton, Bruce, 1947–
  Halley's Comet, 1755-1984.

  Includes indexes.
  1. Halley's comet—Bibliography. I. Title.
Z5154.C4M67 1985   016.5236′4   84-19716
[QB723.H2]
ISBN 0-313-24022-1 (lib. bdg.)

Library of Congress Catalog Card Number: 84-19716
ISBN: 0-313-24022-1

First published in 1985

Greenwood Press
A division of Congressional Information Service, Inc.
88 Post Road West, Westport, Connecticut 06881

Printed in the United States of America

10 9 8 7 6 5 4 3 2 1

To my mother and father
who always encouraged me
to look around, as well as up.

CONTENTS

Illustrations

PREFACE

This book is itself part of the tradition that it examines. Perhaps it would be more logically published after the comet's 1986 apparition, so that all new writing devoted to the comet could be incorporated; this, however, is not practical. Examination of the literature devoted to the last three apparitions of Halley's Comet suggests that after an apparition--very soon after--public, scientific, and scholarly interest wanes considerably. Therefore, this volume is intended to provide a general audience with a socio/biblio-historic insight into the response to the upcoming and previous apparitions; to wait until after the appearance of the 1986 apparition would be self-defeating. This book is, after all, a cultural response (scholarly yet with commercial underpinnings) to Halley's Comet and hopefully will be accepted and used as such.

The time span to which this volume directs itself begins in 1755 with the approach of the comet at the time of its first apparition as Edmund Halley's comet and carries through the first third of 1984. It is our response to the "expected" return of Halley's Comet that occupies this socio-bibliographic study. The focus is not the comet per se, but rather our cultural response to the periodic rising and ebbing of expectation as Halley's Comet comes and goes.

This book was conceived with the multiple purposes of providing basic bibliographic access to the mainstream of English language material about Halley's Comet and at the same time presenting this material in such a way that the reader could have a sense of the chronological continuum of our cultural response to this natural phenomenon. Efforts have been made to keep the text's language as unscientific and nontechnical as possible. However, the reader will recognize that scientific and technical language becomes more prevalent the closer we move to the present. This does not represent an oversight or change of audience, but rather reflects the the remarkably rapid development of high-technology during the last twenty years, affecting the way that we think and speak about, and interact with, our environment.

It is hoped that this bibliographic study will well serve the person merely curious about Halley's Comet, the serious student studying the comet, and the teacher who must teach about the comet itself or who wishes to use the event of the comet's apparition as an entry into the cultural psyche of 1759, 1835, 1910, and/or 1986. In the latter case, this book should have relevance to the disciplines of history, literature, sociology, political science, popular culture, religion, art history, American studies, history of science, physics, and astronomy. It is hoped that those individuals with more sophisticated scientific needs may find this volume in some ways useful as well, although it has not been written with them in mind as the primary audience.

This book has limitations--some of which are self-imposed. Its intent is not to be comprehensive, but rather deliberately selective. An attempt at comprehensiveness in the English language alone would increase the number of entries by at least five times. Every local newspaper in the country could be mined for local response to Halley's Comet on the days of 18 and 19 May 1910 alone. Representative pieces from local newspapers from various geographic locales in the United States are provided, as well as comprehensive coverage of the New York Times, our national newspaper of "record," and the London Times, Great Britain's newspaper of record. So, too, there is fairly substantial coverage of the representative scientific (American and British) and popular journals (mostly American). There is a definite bias toward material originating in the United States. However, coverage of the London Times, publications from the Royal Astronomical Society and the British Astronomical Society, and Nature, the major popular British science magazine, are covered quite thorougly. This stems from the influential relationship English culture has had on its American counterpart--especially during the 1759 and 1835 apparitions of Halley's Comet. One must be mindful that there is, in addition, a mass of material available in other languages which, if addressed, would have further increased the size of the manuscript--this trove I leave to someone else.

A Note on Organization and Style

The bibliography is arranged chronologically, with entries under the same date listed alphabetically by author or title. Entries dated by year only are filed at the beginning of each annual section; entries dated by month and year only are filed at the beginning of the appropriate monthly section, before specific day entries for that month.

Annotations have been constructed to reflect the essence of the material gathered and to contribute insight into the type of cultural response that the material represents within the cumulative context. The annotations attempt, by either quote, near quote, paraphrase, or selective retention of diction to reflect both the tone and style of the material cited. Therefore, unless authorial intrusion clearly suggests otherwise, the reader should consider the entry citation as the source of the annotation.

ACKNOWLEDGMENTS

Among the many who have provided support and aid to me
during my work on this project, I must first thank my wife,
Barbara, and children, Jeremy and Anika, for cheerfully
supporting a man driven--often somewhere else.

My special thanks are also due my colleagues at the
Carleton College Library who have stalwartly carried on in
the face of my being frequently distracted ("cometose"
[sic], they tell me), and during my leaves of absence, and
especially for enduring, with at least mock interest, my
constant chatter about Halley's Comet. An extra expression
of gratitude is in order for the Carleton College Library's
Interlibrary Services Section, which acquired and processed
my needs at what I am told was a record-setting pace.

I must also express my gratitude and appreciation to
the staff of the Carleton College Computer Center,
especially Carl Henry, Les LaCroix, and Cliff Beshers, for
their unflagging support. And again, to my colleague
Richard E. Miller for his invaluable assistance in setting
up computer procedures for data input.

For their willingness to respond to my inquiries I
thank Kathleen Stavec of the New Jersey Historical Society,
Robert Blesse of the University of Nevada-Reno Library,
Barron K. Oder of the University of New Mexico Library,
Conrad F. Weitzel of the Ohio Historical Society, Paul
Eugen Camp of the Florida Historical Society, James L.
Hansen of the Wisconsin State Historical Society, Rijn
Templeton of the University of Iowa Library, Pamela D.
Arceneaux of the Historic New Orleans Collection, Byron E.
Swanson of the Indiana State Library, Sandra E. Fitzgerald
of the **Indianapolis Star/News**, Lyn Stallings of the
Historical Society of Delaware, Ralph Melnick of the
College of Charleston Library, Jackie Pouncy of the State
of Alabama Department of Archives and History, Larry
Jochims of the Kansas State Historical Society, Monica
Rockefeller of the Colorado Historical Society, Ann Graves

of the Texas State Library, Carla Rickerson of the University of Washington Libraries, Michele R. Canney of the **Arizona Daily Star**, Margo Carter of the Oregon Historical Society, Nancy Bartlett of the University of Michigan's Bentley Historical Library, James A. Davis of the State Historical Society of North Dakota, Betty Loudon of the Nebraska State Historical Society, Margot McCain of the Maine Historical Society, Steven R. Wood of the Utah Historical Society, and Sybille Zemitis of the California State Library. I am also most appreciative of the cooperation of **The New York Times Co.**, **The Christian Science Monitor**, **Aviation Week and Space Technology**, The National Aeronautics and Space Administration, the International Halley Watch, and Joseph Laufer, for allowing the reprinting of illustrative material.

Special acknowledgments are due to Kay Schwartau for bringing several Halley items to my attention, to Susan Thurston for her proofreading, to Carleton College faculty, Robert Mathews, Kirk Jeffrey, Robert Wood, and Russell Langworthy, for reading the manuscript and commenting on it at various stages of its development, and to Sally Robinson for producing the final copy of the manuscript. For Sue Swan's expert assistance with the book's illustrations, I am especially grateful.

And finally, my grateful appreciation to Mary Sive, my editor at Greenwood Press, for her unflagging help, and more importantly, for her patience.

Thank you all.

Halley's Comet, 1755-1984

INTRODUCTION

Heavenly bodies have taken their earthly toll. We find ourselves sun-worshipping, sunburned, moonstruck, lunatic, and--occasionally--cometose(?). Be we fools for our mortal fascinations? Just why is it that after the sun and the moon, this hairy star, this flaming sword, this dirty snowball, this Halley's Comet, has so indelibly left its mark on our society? After all the sun and moon are constants in our daily lives, yet Halley's Comet appears to us, on the average, only once every 76 years. Can there be any doubt that this sky-whizzing celestial visitor is, indeed, the greatest comet on earth?

The lore and superstition that have accrued to Halley's Comet in its past appearances have provided those observing each successive apparition with the opportunity to fall prey to the cumulative fear of generations past. Each new generation compounds its own irrational relationship with the natural environment. Our fixation with the association of historical events is a recurring pattern in the literature of Halley's Comet. This, of course, can be attributed to the comet's periodicity, linking both the individual and the society to the past. The watcher becomes fixed, or at least the consciousness does, in the continuum of time.

The comet catches and holds the collective imagination because it is, for most, a once-in-a-lifetime experience. In many ways it is similar to the celebration of a milestone birthday or anniversary, the recurrence of which is noted by extraordinary preparation and celebration. The coming of Halley's Comet is something to which we can look forward or to which we look back, and as such becomes an object of anticipation or reminiscence. And just as the celebration of special anniversaries, the silver or golden for instance, and special birthdays (Sweet Sixteen, twenty-one, or the centenary) have been ritualized, so, too, has observance of Halley's Comet at each of its four appearances since Edmund Halley predicted that it would reappear every 76 years or so.

Comets had long been feared for an array of irrational reasons. But since Halley's computation and prediction published in 1705, stating that the same comet had appeared and would continue to appear periodically, and the subsequent realization of his prediction, the discomfort of the unknown has with each return of the comet been considerably reduced--diminished not by what we know of its physical essence, but by its predictability. **Man** had reasoned its schedule. Each visit of the comet since 1705 has been met by great public anticipation--a fear of the known, if you will. Although such anticipation breeds some anxiety, the prevalent tone has tended rather toward celebration. Each appearance of Halley's Comet is celebrated as a recurring monument to the age of reason--man's reason. What in the early eighteenth century was seen as one of the first empirical vindications of Newton's theories of gravitation, now provides us a bit of cosmic reassurance midst what might otherwise seem a world of natural and social chaos. We celebrate daily with our watches and clocks the regularized comfort of natural order in the regularity of day following night and night day, our rhythmic circadian lives routinized by the observation of Halley's Comet about once every 76 years. In the microcosm of the calendar of our finite lifetimes Halley's Comet is macrocosmic.

The once-in-a-lifetime reappearance of Halley's Comet is certainly one of what Michel Foucault has termed the "big events." Foucault suggests that these "big" events have no inherent "meaning." Instead, people impute meaning to the event, then use the event to validate their own ideas and values--a quite common kind of circularity. Hence a major public event of this sort will attract quite a charge of symbolic meaning; the discussion of the event (in this case, the return of the comet) is helpful because it brings latent beliefs and values out into the open.[1] Therefore we see not only information about the comet's position and physical appearance and substance being communicated, but also, quite unintentionally, general social attitudes about religion, sex, race, ethnocentrism, the unknown, science, technology, literature, art, language-- all things one would not normally associate with a comet. However, the society reveals much about itself through its reaction to Halley's Comet. Furthermore, the periodic reappearance of the comet provides the opportunity to observe the changes in a society at 76-year block intervals. For invariably response to the comet elicits parallel revelations for comparison. With each passing of the comet the earth is a far different place from what it was at the time of the previous apparition. That difference seems to be growing at an exponential rate.[2]

The Spiritual

Since the ancients first conceived of the idea of a deity the awesomeness of this world's natural phenomena has been attributed to the supernatural. In this vein, Halley's Comet has garnered more than its share of attention from the theologically inclined mortal. "Of the many aspects of astronomy, it was comets--representing an event of novelty in an otherwise well-ordered heaven--which exercised the strongest claim upon the popular mind."[3] Comets had been viewed fearfully through the ages and were more often than not associated with whatever bad thing that might have been coincidentally happening concurrent with their appearance. Even the word **disaster** (Latin for "bad star") was originated to describe these events which to the populace so obviously resulted from a comet's appearance. With each new generation the cumulative effect was to instill a negative association to comets--a cultural bias evolved. Indeed, the literature devoted to Halley's Comet is rife with historical allusions to all manner of havoc--floods, droughts, earthquakes, disease, wars fought, battles lost, cities fallen, monarchs expired, or rising expectations that the earth would be destroyed in a sudden ball of cataclysmic fire as it collided with the comet.

One would think that the triumph of logic and mathematical computation in calculating the ellipse of the comet's orbit might have reduced superstition and fear. Ironically, Halley's recognition of this comet's regular return and consequent confirmation of Newton's theories on the laws of physics served not completely to allay man's fear but rather charged his expectation with the highest levels of anxiety.

With its reappearance in 1759 the comet was for the first time **Halley's** comet. In spite of the explanation of the comet in terms of irrefutable mathematical truth and natural law, the common view was that such natural order was imposed upon the universe by a Supreme Governor. God provided man with reason in order that man could begin to fathom the greatness of God's natural order. So just as superstitious forebears, panic-stricken, viewed previous apparitions of Halley's Comet as the accusing forefinger of God pointing at them, the generation of the mid-Eighteenth Century saw the comet as an assertion of God's power to impose a natural order midst what would otherwise be chaos. In both cases a passive mankind stood in awe of an active and awesome supreme being. Man's reason had not yet rationalized him very much distant from those ancients who quaked before the comet because they did not have any understanding of it. Those who expected it in 1759 but understood it no better, quaking before it as a manifestation of the power of a wrathful Old Testament God. The spirit of the Mathers and Jonathan Edwards had not yet faded from the American psyche--and it had been only a century earlier that the Puritans ruled England.

The apparition of 1835 sees the focus of the literature shift to the accomplishment of Halley's having predicted the return of the comet. The triumph of the rational mind is viewed in some Christian quarters as a triumph of intellect that enhances moral destiny in as much as it serves to dampen superstition and false belief. God is set in the role of the power controlling or regulating the universe--not just overseeing it. God is seen as a consistently active presence in the universe rather than merely a passive overseer that occasionally intercedes to assert Himself. Therefore it is man's (Halley's) understanding of the comet, i.e., God's presence as regulator, that is important, rather than a theological fixation on the comet as a flexing of heavenly muscle. Its appearance at the dawning of the Victorian era in England, which coincidentally happened to be the end of the second great spiritual awakening in America, is less symbolic than it is a gauge of the dissipation of spirituality that transpired between 1835 and the close of the Edwardian interlude in 1910.

The appearance of the comet in 1910 was viewed by the educated by and large with a Victorian sensibility that saw in the comet's apparition at the predicted time and place an example of the supreme order that exists even in the apparent chaos of the world. The chance of collision about which there is so much speculation is viewed as merely a question of whether God wills it or not and therefore the chance of the world's destruction is a manifestation of God's power. There is some rationalization, that, indeed, there is no danger of collision at all since there is a supreme benevolent God that will protect man. However, among the less educated classes of the bubbling melting pot that America had become by 1910 it was a far different case. Rumor ran rampant, stoked by a press that sensationalized the comet out of proportion to any kind of objectivity. Consequently, those denominations that tended to emphasize the supernatural in their doctrine, especially the Catholic and Baptist churches as well as fundamentalist sects, found themselves besieged with penitent believers as the comet approached ever closer. It is interesting to note that, for the most part, these Catholics, Baptists, and fundamentalists were recent immigrants, blacks, and poor rural whites respectively. By and large, the activity of white Anglo-Saxon protestant mainstream upper and middle-class America was not reported in the 1910 press. Whether this was because of their abiding faith in the goodness of themselves, their God, or both, remains to be seen. As a matter of fact, it is striking that more often than not the American press is geocentric as well as ethnocentric. Persons who are reported reacting superstitiously or hysterically to the comet are usually described as women, blacks, immigrants, or whites living in a distant place.

With the rise of fundamentalism and its attendant aversion to science and modernism during the first decade of the twentieth century, the advent of the next apparition

of Halley's Comet provided a focal point for reaction to
the scientific/academic community--the focus was a natural,
not supernatural, phenomenon. Ironically, some scholars
chose close scrutiny of Biblical text to corroborate
empirical hypotheses as to the comet's dates of appearance
and orbit, rather than to prove a linkage to God. Biblical
references that have been invoked in the name of Halley's
Comet are II Samuel 2 and I Chronicles 21, both of which
refer to an angel with a fiery sword seen in the heavens by
David who took it as a sign from God to build Solomon's
Temple to house the Ark of the Covenant. Also, scrutiny of
the Talmud reveals reference to a star that reappears every
seventy years.

Science

     The appearance of the comet in 1759 went for the most
part   unnoticed   in   the   popular   press   because its
predictability had not yet caught the popular imagination.
Indeed, the comet was awaited by some as per Halley's
prediction, but it was only half expected. It is evident
from the 1757 **Poor Richard's Almanac** that the comet was
awaited in 1757-58 instead of the 1758-59. That it was
first sighted by an amateur astronomer, Palitzsch, at
Dresden, Germany on Christmas day 1758 was more an accident
than anything else.
     The apparition of 1759 was the first appearance after
Halley's prediction and its actualization was a great
confirmation of Newtonian physics. "What atomic physics is
to our own century Newtonian astronomy was to the
eighteenth. In Europe the best scientific minds vied with
each other to extend and perfect Newton's mathematical
demonstration of the solar system."[4] The scheduled
reappearance of what then became known at once and forever
as Halley's Comet, immediately elevated cometary astronomy
to a state of perfection comparable with that of the other
branches of astronomy.[5] In fact, the comet's appearance
in December 1758 served to enhance the entire discipline of
astronomy, helping it gain a parity of esteem with the
other scientific disciplines. As such the apparition is
one of what Michel Foucault calls the "thresholds"
suspending the continous accumulation of knowledge (and in
this case, superstition), interrupting its slow
development, forcing us to enter a new time, cleansing it
of all imaginary complicities. Such an event directs us
toward a new type of rationality and its various
effects.[6]
     The immediate response to the comet's appearance in
1759 was not one great scientific engagement, however.
Rather the literature exhibits a pedestrian bias for
relative placement of the comet in the heavens and visual
description. Since the Royal Astronomical Society was not
formed until 1820 none of its various publications existed
as a forum for discussion and reporting about Halley's

Comet at its first predicted return. And in colonial America, of course, there were few serious amateur astronomers. For the most part, British observers passively described instead of allowing the intellect to actively engage the comet. By and large, it was society's belief in its God that engaged the comet, and not its belief in its own analytical abilities. It was science at its least imaginative. The case was somewhat otherwise on the European continent where in Germany and France mathematical observation was the rule and as such was reflected in the literature.

The minimal response to the 1835 apparition from America is notable. The nation was fifty years young, very much asserting itself commercially, industrially, technologically, culturally in philosophy, literature, and language, and was expanding its borders. Yet in the realm of science America remained, relatively, a backwater to the activity taking place in the major European centers. In 1835 there were few astronomers (no more than 15) [7] and no observatories per se [8] in the United States. The procurement from Europe of the minimally necessary tools of astronomy, the transit instrument and a clock, was beyond the means of most would-be astronomers. And an adequate telescope required a sizable capital investment.[9] The finest telescope in America at the time of Halley's 1835 return was a 5-inch Dolland refractor at Yale, purchased in 1828 with a gift from an Oxford, Connecticut farmer, of 1200 dollars but no observatory was built to house it.[10]

It is not surprising, then, that most of the literature devoted to the comet in its second post-Halley apparition appeared in European scientific journals and newspapers. Those articles that did appear in the American press were often reprints or excerpts of what had already appeared in Europe, and only 53 articles in the entire field of astronomy were published in American journals during the entire 1830s.[11] It is no wonder that writing about the comet was dominated by the Europeans. Yet, it is equally apparent that what little serious astronomical activity there was in the United States during the first half of the decade significantly increased during the years 1835-39. There was a 58 percent increase in the number of persons involved in astronomical research and a 200 percent increase in the number of articles published.[12] In both cases this represents the most real quantitative progress between any five-year cohort in the period 1815-1844. It would be difficult not to attribute these facts, at least in part, to the scientific and popular interest generated by Halley's Comet. In the surge of democratization taking place in Jacksonian America science too became popularized, and was no longer the realm of the gentlemen philosophers able to combine science with a life of leisure. Astronomy among the various scientific disciplines was most appreciated by the common man because one needed neither elaborate instruments nor a higher education in order to pause with wonder at the workings of the heavens--one needed only to look up. In addition to accessibility, astronomy invited public interest in that the events of the heavens had long been an inspiration for poetry, folklore, and viewed to be a prima-facie theological proof.[13] Unlike the physical phenomena of the other natural sciences, the public had long since been acculturated to those of the sky.

A striking characteristic of the Halley's Comet scientific literature in 1835 on both sides of the Atlantic is the prevalence of the word **about** as a quantitative adjective. It is evident that for most observers observation was still, at best, educated guesswork. The literature is dominated by articles or reports describing the comet in terms of its location in relation to other stars. The scientific literature of the period obviously was written to facilitate further observation through the diffusion of what was already known rather than to advance science.[14] Later, beginning in 1837, lengthier reports of observation appeared--a pattern to be repeated in 1910. Articles that appeared in popular journals were devoted, for the most part, to the history and lore associated with Halley's Comet. The telescope, sextant, and human eye were the primary methods of scientific engagement.

Since in 1759 the United States had been a British colony, 1835 is the first opportunity to evidence American nationalism vis-a-vis the comet. The overt rivalries rooted in the American Revolution and the War of 1812 were still fresh in public memory--many veterans of those conflicts still lived. In the ferment that was Jacksonian America there was a conscious effort being made to be American, that is un-British. There was a budding sense of American nationalism in politics, language, literature, trade, etc. So it is not surprising to find that the handful of astronomers practicing their science in the United States also felt competitive with the British. "Since Newton's day, astronomy had been the most elegant of sciences, and a nation's contributions to that science were commonly regarded as a sure index of cultural achievement. Sensitive to the jibes of foreign critics, some Americans fostered astronomy as a matter of national pride. Nothing, wrote a fawning Edward Everett in 1838, 'would be half so acceptable to the men of Science in Europe' as for the United States to erect an observatory."[15]

There existed among America's few practicing astronomers an intense desire to be the first to sight Halley's Comet. After all, the world was expecting it this time. Yale was the only place in America in 1835 that had a telescope with which accurate readings of heavenly bodies could be made, and made they were. However, the work of Olmsted and Loomis at Yale was largely neglected by the European astronomical community. The intense desire to beat the British in first sighting the comet is apparent in Olmsted's and Loomis' erroneous claim to have first sighted the comet on 31 August 1835 at Yale, when one week earlier a sighting is noted in the **London Times**. Certainly the presumptuous error of the Yale astronomers' claim must be attributed, in part, to the slow communications of the era.

The point not to be lost, however, is that they very much
wished to be first. Ironically, the British were not the
first either. That honor belonged to the astronomer of the
Collegio d' Roma, in Rome, Italy. The Americans must have
felt some satisfaction in that, at least, it was not the
British. To be ignored by the British was one thing, but
to be both ignored and beaten by them would have been quite
another.

The scientists of 1835, with a distinctly
pre-Darwinian perspective, viewed Halley's Comet as an old
comet in that it had been observed for over five hundred
years. This perspective reveals a myopic sense of time
that would be broadened with the advance of science. At
the next apparition, in 1910, post-Darwinian era scientists
recognizing that the comet had been observed for over two
thousand years came to exactly the opposite conclusion,
stating that the comet must be young in order still to be
observable with the naked eye without having sufficiently
dimmed to make only telescopic observation possible. Like
any object or phenomenon, the comet is viewed from a
particular point in time, and often the insight gained is
not of the object or phenomenon but rather of the viewer.

In the years and then months leading up to the 1910
apparition it becomes clear that the scientific community
had developed a schedule fixation in regard to the comet.
The goal became to derive a timetable that would consider
speed, path, and time of appearance. Predictability was to
be the proof of scientific achievement. The culmination of
this attitude was the conferring of honorary degrees upon
Cowell and Crommelin by Oxford University for their work in
determining the exact date of the return to perihelion in
1910 (see entries 156, 297, and 353). The honoring of the
two is at once a tribute to their own ingenuity and
competence and an indictment of the inexactitude of science
at that time--that correct computation was deserving of
such note. Their accomplishment, by no means meager,
pales, however, next to the pioneering computations and
prediction of Newton and Halley almost 150 years before.
The state of astronomical science was in transition from a
body of knowledge and method that devoted most of its time
and energy trying to refine accepted premise and
methodology to a new wave that was beginning to embrace
technological tools in order to advance the frontiers of
knowledge.

For the first time significantly different methods of
astronomical observation would be turned on Halley's Comet.
Since its last appearance in 1835, photography (the
photo-dry plate), spectrography, photometry, and
polariscopy had been developed. These observational tools
enhanced scientists' ability to calculate orbit, measure
brightness, as well as the length and the composition of
the comet's tail. We might expect that just as the advent
of astro-photography in the late nineteenth century
provided a quantum leap as a powerful new tool with which
the scientist could engage the phenomena of the heavens in
1910, the development of electronic and rocket technology

during the twentieth century will have an even more important impact on scientific and, for that matter, non-scientific observation of Halley's Comet in 1986.

The worldwide rivalry that developed among astronomers in 1835 to be first to sight the comet was repeated with even more intensity in 1910. In 1908 the American Astronomical and Astrophysical Association appointed a committee on comets to determine what kinds of research should be done, its focus and its methodology (see entry 204). In parallel fashion the British Astronomical Association went ahead and developed its own plans for observation. Curiously, the British later complained about not being consulted by the American association when the Americans compiled and distributed a circular on how best to photograph Halley's Comet. It was apparent that the Americans' initiative was viewed by the British astronomical community with condescension and at the same time as a brazen threat to their own perception of British scientific preeminence. At any rate, there was little international cooperation. Perhaps this was just another manifestion of the strong forces of nationalism and isolationism of the era immediately preceeding World War I.

The scientific rhetoric of 1910 articulated that "the comet will be the most striking phenomenon of its kind during the present generation." It is noteworthy how similar are the statements of scientists in the 1980s. Indeed, much of Halley's Comet's allure is its regular recurrent singularity. The desire to place previous apparitions of the comet using mathematical calculation was, of course, motivated by scientists' desire to learn more about the motions of the comet during its present apparition. However, the non-scientific community used the dating of the comet's appearances in a way not unlike that of the genealogist—it was a means by which people could place the comet and, by association, themselves and their forebears.

Despite scientific assurances that the comet was harmless, there was nevertheless widespread fear of it. The unwillingness of the general public to accept the authority of the scientific community was evidence of a popular distrust of science and scientists. Indeed, the seeming inability of scientists to agree in the press and the professional literature on the composition of the comet's tail or whether the earth had passed through the comet's tail, did little to instill public confidence. Once again, the authorities in attempting to allay fears of disaster, reinforced the public's suspicions that there must be at least a chance of harm if so much attention was being devoted to its unlikelihood. The high profile of the scientific community in the popular literature and press during 1909 and 1910 served to accentuate for the public the disagreement that existed among the so-called experts. This fact was aptly put by a contemporary observer, "To future generations it will seem as if Halley's Comet in 1910 had fallen among astronomic experts like the proverbial bone of contention among ravenous dogs."[16] Yet another observer of the scene suggested, "the only fear that need be felt is that after such elaborate preparation by astronomers the comet may not do as they expect."[17]

Superstition in the guise of science was fairly prevalent. There was a widespread belief, for instance, that air could be bottled as the earth passed through the comet's tail and that it then could be scientifically assayed in order to determine the composition of Halley's Comet. However, this scheme flies in the face of the known density of the earth's atmosphere and the acknowledged inability of the rarified gases of the tail to penetrate it. Publications giving attention to both of these propositions were widespread and yet failed to point out the obvious contradiction between aspiration and fact. The failure of the scientific community to capture a sample of the comet was seen by some as the greatest disappointment of the entire 1910 apparition. The bottom line is that there was insufficient data to determine the effect of Halley's Comet. This is borne out by statements in the literature of the 1980s to the effect that we know very little about comets. The desire to capture a sample of the comet and examine it remains a prime objective of plans for the 1986 observation of comet and an aggregate of billions of dollars, yen, rubles, marks, francs and pounds will be spent in the attempt to realize man's desire to reach out and touch Halley's Comet.

In spite of what was a generally passive attitude toward the comet that will be discussed later herein, there were many active attempts to observe and photograph it from balloons at altitudes of two to three miles. The were even some attempts to coordinate balloon observations at several locations. However, more often than not, aeronautic observation was merely a matter of expediency in the face of inclement weather. Serious scientific vehicular observation would have to wait for the comet's next visit in 1986.

The literature of the 1986 apparition begins, for practical purposes, in 1967 with a study by Lockheed Missile Corporation that shows that a space probe of Halley's Comet is feasible (see entry 1014). Such a probe was attractive even nineteen years prior to the expected perihelion of the comet because of its size and brightness, and most of all, its predictability. Because of the obvious expense in mounting a space probe, much of the literature that prefaces the 1986 apparition dwells on money. Since this would be the first cometary probe ever it would be necessary to produce heretofore non-existent equipment instead of relying on in-place instruments as had been done with previous apparitions. At earlier appearances of Halley's Comet man responded to his visual (either naked eye or telescopic) perceptions of the comet. Now with the electronic computer on the scene, the 1986 apparition will find us responding not to our own immediate perceptions of the comet, but rather to a machine's response. It is ironical that this one step removal of immediacy in the interactive relationship between comet and

12

man, will provide man with much more information, paradoxically bringing him nearer to Halley's Comet than ever before possible. The reality of such mechanical intermediaries seems to suggest an affirmation of Max Frisch's wry observation that technology is the knack of arranging the world so that we do not have to experience it.[18] Of course we shall experience it, but it will be from a distance once removed--immediacy is replaced by intermediacy.

In the United States there has been a protracted struggle to get funding appropriated for the National Aeronautics and Space Administration (NASA) so it could mount a Halley's Comet space probe. The lack of success in achieving this goal, even with intense lobbying, points to the constant tension that now exists between the scientific community in its reliance on government funds to meet the great expenses incurred in developing the sophisticated technology necessary to today's scientific research, and the government which dispenses those funds based on what is often a completely different set of priorities. The failure of the United States to mount a Halley mission and the attendant debate and political process in reaching that decision, exemplify the degree to which complicated and expensive technology is the basis of today's scientific research, and as a consequence how the scientific community has become reliant on government subsidy. Nowhere in the literature of 1759, 1835, or 1910 was this apparent. The government's seeming insensitivity to scientists' priorities is based not on a hostility to science per se, but rather on a different perspective of what is cost/beneficial in terms of policy and program--i.e., politics. In this case, the government views the focus of the space program to be the Space Shuttle because it has the possibility of immediate military and commercial benefits, whereas space exploration is purely speculative in terms of return on investment. The government hopes to demonstrate the flexibility and economy of the space shuttle and at the same time make political hay by having it conduct in-flight experiments on Halley's Comet in 1986.

The politics of Halley's Comet dominates the literature of 1975-82. Domestically, it is the debate over money and concomitant issues of the value of pure versus applied research and national prestige. On the international scene the difficulty was to get normally distrustful national governments to agree to coordinate observation and to freely exchange scientific data via the auspices of International Halley Watch. The International Halley Watch represents an effort by scientists, not governments, to foster scientific cooperation on the international level (instead of on the national level as was the case in 1910). The possibility to undertake this successfully is a direct result of advances in communication and travel technology. Plans and findings can be communicated instantaneously to anywhere in the world and scientists can travel in a day's time to almost any meeting in the world.

13

With several nations having made the commitment to send spacecraft to rendezvous with Halley's Comet, it is not surprising to find that recent scientific literature devotes much attention to the building, outfitting, propelling, and navigation of the rendezvous spacecraft. This is, indeed, an ironical inversion of the attention focused in 1910 on calculating the course of the comet toward a visual rendezvous with the human eye (in terms of observation). In 1910 some articles about the observation of Halley's Comet, of course, dealt with the equipment used for observation. Similarly, in the 1980s many articles deal with either the spacecraft itself or the instrumentation on board. Much of this literature is not about the comet per se, but rather about the technology that will be applied to it. Because in the last seventy-five years technology has become, obviously, so prevalent and complex and is in a constant state of development, it dominates discussion at the expense of the comet. One would expect that once rendezvous is made with the comet the focus will shift to the comet itself. Yet, the level of attention given in the literature to method in preparation for the comet's appearance is another important difference between now and the last apparition. The scientist of 1986 is so much more dependent on technology than had been his counterpart of 1910, but then, of course, the scientist of 1910 was much more dependent on technology than had been his counterpart of 1835, and so on. This is not at all surprising. The increase in the number of scientific journals, the advance of available technology for experimentation, and the ease of daily communication between scientists, have all served to encourage today's astronomers and engineers to report what they are **doing** rather than what they have done. Furthermore, the fragmentation and specialization of science into subdisciplines, combined with the ease of travel over great distances, has caused the literature devoted to the comet to reflect many specialized conference proceedings.

Popular Response

The comet's 1759 return was awaited by a relative few who remembered Halley's prediction, and even some of these waited with skepticism. The combination of uncertain expectation and the absence of any group or body (the Royal Astronomical Society was not founded until 1820) to marshall interest in Halley's Comet served in 1759 to lessen what we have come to expect in the way of public response to the comet. In any case, English language response to the appearance of Halley's Comet in 1759 was scant. Newspapers of the day succinctly reported the comet's appearance and location to a large and generally literate audience. All in all, there was surprisingly little attention paid to the comet during its first go-around as Halley's Comet.[19] The public's collective imagination had not yet been aroused by its periodicity. For that matter, in the public mind, until the comet actually had appeared in 1759 there was no certitude of its periodicity--there was no **Halley's** comet. The comet caught the public imagination only after its appearance verified Halley's hypothesis, thereby becoming a full-blown part of our scientific **id** and social mythology.

The English language literature devoted to the comet in 1759 was predominately British. The popular press found it much more economical to publish in Great Britain where there were far more potential readers. And since the American colonists were at this time fighting as loyal subjects alongside the British against the French, maintained a keen interest in affairs of British society and state and avidly awaited British popular magazines and newspapers. It is not surprising, then, that what articles did appear in the American press were, for the most part, reprints of those already having appeared in England.

Since the correctness and implications of Halley's calculated prediction were not yet fully accepted or understood by the general populace, there existed still a great deal of superstition about and fear of the comet. Such concern is clearly manifested in the press's attempts to allay the public's fears of a collision between the earth and the comet. However, the effect was the opposite of the press's intent. Each refutation strengthened the convictions of the fearful that there must be, indeed, something of potential danger if so much attention were being given to the small probability of such an event taking place. Attempts to minimize public concerns were also undermined by the very same press publishing accounts of calamities, disasters, victories, and defeats that have been associated historically, by pseudo-attribution, with the comet. The expectation of the public therefore was raised to expect a similar situation with the current apparition; and sure enough, if one is so inclined, there is the victory of Wolfe at Quebec, the fall of Quebec, the death of Spain's Ferdinand VI, and more. It is interesting that people would think that victories are won, battles lost, or monarchs die only every 76 years. Coincidence, in the case of Halley's Comet, is the child of selective imagination.

In 1835, as was the case in 1759, much of what appeared in the American journals and press was reprinted from what had already appeared in England. The reason for this had changed, however. The American colonies were colonies no longer; they were now the United States of America and as such were in 1835 experiencing a generally anti-British cultural tone--it was, after all, only 54 years after the American Revolutionary War and 20 years after the War of 1812. There is a discernable contrast between coverage of the comet in Dickensian England and Jacksonian America. The United States with a much smaller population, had a literacy rate of approximately 78 percent. In England, with a much larger population, the literacy rate was considerably lower, approximately 65 per cent, although the literate population numbered more than in the United States.[20] The regular coverage of the 1835 comet in the **Times** (London) suggests a popular appreciation of the phenomenon that was not yet apparent in America. So much so that a reader of the **Times** (London) in a letter to the editor asks an "astronomically oriented" correspondent to write his **Times** pieces in non-technical English so as to be more easily understood by the average reader. Such popular appeal in England is not surprising in view of the institutionalization of astronomy only 15 years earlier in the founding of the Royal Astronomical Society (1820). On the other hand, in 1835 the United States did not even come close to having a world-class observatory and only one telescope (Yale's) was capable of truly state-of-the-art observation.

Jacksonian egalitarian democracy was prevailing in the political sphere and there were parallel cultural and social forces at work. There was a concern for things distinctly American. As with the previous apparition there was a preoccupation with the chance for collision with the comet, but in 1835 the residents of the United States found themselves exhibiting an idiosyncratic American brand of concern. The fascination with the prospect of destruction curiously appealed to the Puritan/Calvinist strain in American culture. A fiery collision with the comet could be viewed as an incarnation of hell--a cataclysmic end of the world that was consistent with the tide of spiritual awakening that was awash in American society during the decade leading up to the 1835 appearance of Halley's Comet. The publication of articles dwelling on the potential for a collision of the earth and Halley's Comet had the psychological effect of self-fulfillment. People who would never had given the possibility any thought at all became more conscious of such a possibility with each reading. The possibility was given credibility by its mere appearance in print, and more so with each reappearance.

The public uncertainty and anxiety about the comet presented commercial opportunity. 1835 saw the first books about the comet aimed at a popular market. For one such monograph the intended market was children, whose fears were to be rationally allayed. Two others, targeting the general adult market were conversely filled with dire predictions of natural disasters caused by cometary influence--earthquakes, intense summer heat, crop loss, etc. The effect was one of mutual cancellation. No matter what rational explanations of the comet were offered to children, the child would be in the end influenced by the adult view. Simultaneous with the growth of public interest in the appearance of Halley's Comet is growing evidence of public skepticism over the obvious disagreement among practitioners of what was perceived to be such a precise science. Published accounts provided the average reader with the dilemma of deciding whom among the experts to believe, if anyone at all. Scientific objectivity was nothing more than method to the public--the method had not yet come to reflect result.

The American press was ready and waiting for Halley's Comet in 1910. Although there was great attention given to the comet on both sides of the Atlantic, the United States by 1910 had developed a strong national press, both newspapers and popular magazines were abundant, and through the emulation of British professional societies, so too were professional scientific journals. Coverage was complete and generally originated in the country of publication, the exception being many reports from stations in the British empire having appeared in the English journals. American reaction to the comet in 1910 clearly demonstrates that since 1835 the United States had matured as a recorder and debater of its own cultural life.

In spite of low-brow celebration of the comet through heavy use in advertisements, as a theme for parties, and as an appellation for mixed drinks--the **comet cocktail, cyanogen cocktail**, and the **syzygy fizz**, the public was not completely disinterested in the higher scientific aspects of Halley's Comet. However, for all the advances made in the sciences between 1835 and 1910, the general public in 1910 still viewed the differing scientific pronouncements on the subject of Halley's Comet with skepticism. The scientists had entrenched themselves at the nation's universities and from that position in 1910 either offered themselves or were called upon as pundits by both the local and national press. The 1910 appearance of Halley's Comet probably did more than any single scientific event to that date to put the professional scientific community before the public eye. There were those among the scientists that rather enjoyed the role of public pundit and allowed themselves to be manipulated by a national press that was more interested in selling newspapers than it was in necessarily getting the facts right. **The New York Times** is a perfect example of the press's using the facts as a counterpoint to outrageous speculation. As the comet approached perihelion and transit the **Times** seemed to alternate stories of poison gas and persons who broke down in utter fear with scientific pronouncements disclaiming the need to worry. The result was a crescendo of public anxiety and a crisis of public confidence in the scientist. Everytime the scientist reassured the public, there was yet another story presenting an alternate scientific view and then more reports of superstition past and present. The scientist, indeed, perhaps for the first time, became a full-fledged participant in a sustained public discussion of national scope. And, yet, as such was unable to gain esteem for the profession; the scientist remained a parvenu.

As the 1985/86 appearance of Halley's Comet approaches the scientist is firmly ensconced as our society's engineer. The ambiguity of the word "engineer" is in this case most appropriate since the scientist is at once steering our course to the future, intimately involved in governmental policy making, and simultaneously planning, inventing, discovering, testing, and building the technology that will carry us there.

Popular response to the comet in the 1980s so far has been limited to sporadic articles focusing on associating the comet with historical events, much ado about the failure of the United States to mount a Halley's Comet mission, and entrepeneurial effort to hawk Halley's Comet bumper stickers, buttons, and T-shirts (see entries 1217, 1278A, and 1290). The commercialization of the comet will likely reach heights unthought of by the pill pushers, snake oil salesmen, advertisers, and souvenir makers who saw opportunity in the comet seventy-six years ago. When the comet is nearer to earth and the various rendezvous spacecraft prepare to encounter their target international interest will peak. Not only can we anticipate that the press will address the comet and its history with increasing frequency, but there will also be close attention paid to the success of the several space probes. In fact, the electronic sensing devices aboard the spacecraft will relay data back to earth and we can expect to have photographs and other electronic images of the comet transmitted into the home via television with relative immediacy. One expects that immediacy and intimacy will combine to substantially diminish supersititious reactions, at least in the industrialized world. Science acts as the great equalizer, on one hand solving yesterday's mysteries while at the same moment pushing the intellect to question new mysteries discovered on the frontiers of scientific exploration.

Obviously, to the astronomer Halley's Comet is something very special in that it exhibits physical attributes worthy of scientific attention. Yet, it is somewhat more difficult to explain the continuing public infatuation with this particular comet. Certainly, the appearance of Halley's Comet every 76 or so years is not made such a fuss about because it is the only opportunity for John Doe to see an astronomical phenomenon of this kind. Meteors, shooting stars, and comets are relatively common astronomical occurrences. Why then, is it this particular comet that makes us irrational in our desire to be rational? It is because the amount and the nature of the attention that we give the comet has cumulatively imbedded it in our cultural psyche as an archetype. Say comet to someone and the chances are very good that the associative reply will be **Halley's**. Stories of having seen the comet have been passed down from generation to generation, and each generation looks forward to getting its own bragging rights.

Through its various apparitions the comet has served as an inspiration to the artist-- it is the subject of paintings and poems, and in 1910 scaled the heights of Tin Pan Alley, inspiring a song in the Ziegfeld Follies. During the 1950s the comet's fame served as sort of a pop objective correlative in that one of the first "hit" rock-and-roll groups of that era appropriated an association with the comet, playing on their leader's name. Bill Haley and the Comets, gained popularity with their hit song, "Rock Around the Clock" (the song became the first rock million-seller in England and was the theme for the movie "Blackboard Jungle"), and became the the de facto rock anthem. It further imprinted the comet on the public mind--and in so doing probably reinforced the mispronunciation of Halley (as in "alley") as Haley (as in "Bailey"). There is no doubt that a new generation of foot-tapping comet gazers are ready and waiting.

We can be sure that as the comet approaches there once again will come a flood of new advertising, songs, poetry, books (**mea culpa**) extolling the brightness, the magnificence, and/or the importance of Halley's Comet. There is also certainty that in the aftermath of the comet's passing there will be a surge of scientific literature reporting and interpreting the data collected from earth and in situ observation. However, it will probably be different this time. In the past the comet was anthropomorphized and was viewed as an active and threatening agent by a passive populace. In 1986 one would presume that a scientific objectivity toward a natural phenomenon will be the prevailing attitude; and the earth will be generally perceived as having an active rather than passive attitude, sending spacecraft to probe the comet's essence. It is just as certain that popular press and television coverage of the comet will abruptly wane to lie dormant for another seventy years or so.

Politics

At the time of the 1759 apparition of Halley's Comet England and its American colonies were at war with the French and their Indian allies. No one really expected the comet to appear as Halley predicted. Those that awaited it were an eccentric few. The mind of both public and state was turned to terrestrial matters such as the exploits of Rodgers' Rangers and the successes of Generals Amherst and Wolfe. Halley's Comet, at least figuratively, was beneath politics.

Seventy-six years later the comet was awaited by gentlemen and academics, mostly in England. There was little if any government support for astronomy in the United States. President Andrew Jackson was engaged in continual struggle with the Congress. There still exsisted a residue of tension from the struggle over the national bank question, but the hottest issue of the day was what it was to be for the next twenty-five years--slavery. The abolitionist movement was beginning to assert itself and almost every newspaper and popular journal in the United States was affected by the topic. The drums of war beat in the Florida Everglades as the Seminoles resisted eviction. News of Americans being massacred obscured the celestial light show, and as the comet began its movement away from earth Santa Ana counted bodies at the Alamo. Halley's Comet in 1835/36 was both figuratively and literally above politics.

Having been bracketed by the politics of the times during its 1759 and 1835 apparitions Halley's Comet in 1910 itself became a political issue. The age-old association of the comet with the death of a monarch was widely invoked at the passing of England's King Edward VII on the eve of the earth's transit of the comet's tail. In the international sphere, British colonial authorities expressed concern that the superstitious nature of their colonial subjects might be used against the empire, the comet used by revolutionaries as a favorable omen. In China, Christian tract societies were quite active in producing handbills and posters explaining the comet as a natural phenomenon and entreating the populace to remain calm. Authorities in Constantinople took advantage of local inhabitants' fear of the comet to enforce the city's unpopular stray dog ordinance. On the night of the earth's expected transit through the tail of the comet most residents of the city remained inside their homes awaiting disaster while stray dogs were rounded up throughout the night. In the United States Halley's Comet saw heavy duty use as a metaphor in political cartoons, with President Taft, Theodore Roosevelt, and the Republican party bearing the brunt of the commentary. Some persons were so emboldened or deranged by the comet's presence that they suggested that primary elections in Texas and Ohio would be affected by Halley's Comet.

International geopolitics played little role in the scientific observation and reporting of Halley's Comet in 1910, a fact which, to a great extent, was a function of rather primitive communications and travel by today's standards. Yet, the world tides of nationalism and isolationism that eventually led to the conflagration of World War I four years latter cannot be discounted as a real deterrent to effective international scientific coordination and cooperation. Then there was professional politics. For the first time an apparition of Halley's Comet was greeted by large and active professional scientific communities in both Great Britain and the United States. This situation brought about the playing out of hidden agenda--a struggle between British and American astronomical communities for leadership in the observation of the comet (as it turned out the weather played favorites, giving the U.S. a clear edge) as well as the right to claim preeminence in the astronomical sciences in general. In America, there was rivalry among observatories for the prestige of being the first to make discoveries, capture the best photographs, and get the best press. Likewise, there seemed to be a de facto rivalry among the astronomers of the University of Chicago and those of Harvard in vying for exposure as authorities. As often is the case in academe, it was a matter of ego tempered by the practicalities of professional politics, and vice versa. While most astronomers diligently looked skyward a few looked earthward for professional opportunity. Some punditry was as much for self-advancement as it was for the advancement of science, the newspapers of the day playing upon academic vanities for their own gain. Among the press **The New York Times** made it quite clear that it viewed itself as the "chief comet organ" of the nation. **The Times**'s hiring of an astronomer, Mary Proctor, as its special correspondent for the apparition of Halley's Comet has to be one of the earlier instances of the American press engaging a scientist on a sustained basis to report specially on a scientific subject, and was to some extent precedent setting.

The politics of the 1985/86 apparition is very much different than that of 1910. The development of electronic communications and supersonic travel have served to make the world a smaller place--McLuhan's global village. Because the science of astronomy has become so fragmented with specialization, it is imperative that the various specialists communicate their theories and discoveries to colleagues working in other areas of the discipline. Such interdependence mitigates against the negative rivalries that one saw during 1909 and 1910. The technology necessary to conduct astronomical science today (traditional observation, deep space observation, and in situ observation) is very expensive. This reality alone fosters cooperation among astronomers since there are few observatories, if any, anywhere in the world which have on site all the equipment necessary for the kind of comprehensive observation and experimentation being planned for Halley's Comet. On each of the spacecraft which are being sent to meet Halley's Comet there are experiments from scientists of other nations.

So even in the face of almost constant international political tension, east-west and capitalist-communist, barriers are being circumvented in the name of Halley's Comet. It is apparent that the desire to know more about Halley's Comet is stronger than the impulse to prevent an adversary from knowing about it. The International Halley Watch, an organization created by scientists, not governments, to coordinate and facilitate observation of the comet at its 1985/86 apparition has been joined by over 800 scientists from 47 countries, including the earth's military and political superpowers. It seems that in a political context Halley's Comet serves as a catalyst, in that "it is obvious today that science and technology, and the various forms of art, all unite humanity in a single and interconnected system. As science progresses, the world-wide cooperation of scientists and technologists becomes more and more of a special and distinct intellectual community of friendship, in which, in place of antagonism, there is growing up a mutually advantageous sharing of work, a coordination of efforts, a common language for the exchange of information and a solidarity, which are, in many cases, independent of the social and political differences of individual states."[21]

It is finally apparent that Halley's Comet returns not to destroy, but to fulfill humankind's urge to know. It was that urge to know that gave it its name, and which gives us a continuing legacy that shines in our eyes and in

our imagination. Halley's comet is the greatest comet on
earth.

NOTES

1. Michel Foucault, **The Archeology of Knowledge**, Tr. A.A.
Sheridan Smith (New York: Pantheon Books, 1972), p. 4.

2. See entry 1249A.

3. Donald Zochert, "Science and the Common Man in
Ante-Bellum America," in **Science in American Since 1820**
(New York: Science History Publications, 1976), p. 8.

4. John C. Greene, "Some Aspects of American Astronomy
1750-1815," in **Early American Science,** ed. Brooke Hindle
(New York: Science History Publications, 1976), p. 181.

5. Amedee Guillemin, **The Heavens: An Illustrated Handbook
of Popular Astronomy** 4th ed. (London: Richard Bentley,
1871), p. 244.

6. Foucault, p. 4.

7. George H. Daniels, **American Science in the Age of
Jackson** (New York: Columbia University Press, 1968), p.
22.

8. David F. Musto, "A Survey of the American Observatory
Movement, 1800-1850," **Vistas in Astronomy**, 9 (1968), 89.

9. Howard S. Miller, "Science and Private Agencies." in
Science and Society in the United States, ed. David D.
Van Tassell and Michael G. Hall (Homewood, Illinois:
Dorsey Press, 1966), p. 202.

10. Bessie Zaban Jones and Lyle Gifford Boyd, **The Harvard
College Observatory: The First Four Directorships,
1839-1919** (Cambridge, Massachusetts: Harvard University
Press, 1971), p. 37.

11. Daniels, **American Science in the Age of Jackson,** p.
23.

12. **Ibid.**, pp. 22-23.

13. Zochert, p. 8.

14. George H. Daniels, **Science in American Society: A
Social History** (New York: Knopf, 1971), p. 173.

15. Miller, p. 203.

16. See entry 436.

17. See entry 458.

18. Max Frisch, **Homo Faber** (New York: Harcourt, Brace, and Jovanovich, 1959), p. 178.

19. See entry 121.

20. See **Historical Statistics of the United States: Colonial Times to 1970**. Washington, D.C.: U.S. Bureau of the Census, 1975, p. 365 (non-slave population based on 1840 census of persons over 20 years of age) and R.S. Schofield, "Dimensions of Illiteracy in England 1750-1850." **Literacy and Social Development In the West: A Reader.** Harvey J. Graff, ed. (New York: Cambridge University Press, 1981), p. 211.

21. Zhores A. Medvedev, **The Medvedev Papers.** Tr. Vera Rich. (London: Macmillan and St. Martin's Press, 1971), p. 3.

Halley's Comet
Perihelion Dates

29 July 2061	2 October 684
9 February 1986	15 March 607
20 April 1910	27 September 530
16 November 1835	28 June 451
13 March 1759	16 February 374
15 September 1682	20 April 295
27 October 1607	17 May 218
26 August 1531	22 March 141
9 June 1456	25 January 66
10 November 1378	11 October 11 B.C.
25 October 1301	6 August 86 B.C.
28 September 1222	12 November 163 B.C.
18 April 1145	25 May 239 B.C.
20 March 1066	8 September 314 B.C.
5 September 989	14 September 390 B.C.
18 July 912	18 July 465 B.C.
28 February 837	10 May 539 B.C.
20 May 760	28 July 615 B.C.
	22 January 689 B.C.

This table is adapted from Yeomans and Kiang (entry 1200) and Yeomans (entry 1052) as presented in Edberg (entry 1270). The Julian calendar is used for dates earlier than 1607.

ANNOTATED BIBLIOGRAPHY

1755

1. Wesley, John. **Serious Thoughts Occasioned by the Earthquake at Lisbon.** London, 1755, pp. 21-22.

Halley predicted the return of the great comet in 1758. This time the comet will move in same line as earth's orbit. Wesley wonders what will be the consequence of a collision between the earth and the comet. His determination is that the earth would be set on fire and burned to a coal. He worries what would be the consequences if the comet is now nearer than Halley predicted?

1756

2. Franklin, Benjamin. "Of the Expected Comet." **Poor Richard Improved:** ... **For the Year 1757.** Philadelphia: B. Franklin and D. Hall, 1756. p. E3 [verso].

Halley's Comet was expected to return in 1757-58 instead of 1758-59 as it actually did. "As these huge tremendous bodies travel through our system, they seem fitted to produce great changes in it." Franklin speculates on the results of a collision between the earth and a comet.

3. Claudius. "Halley and Newton on the Comet Expected in 1758." **Gentleman's Magazine,** 26 (January 1756), 24-27.

The "comet having of late been a prevailing topic of most private as well as public conversations ..." the author thought it expedient for the quieting of uneasy minds to collect from writings of Halley and Newton whatever related to periodic comets.

4. Gemsege, Paul. "Cruelty of Terrifying Weak Minds with Groundless Pains." **Gentleman's Magazine**, 26 (February 1756), 71-72.

   Letter. Refutes pamphlet, the premise of which was the conjecture that Halley's Comet would collide with the earth setting it afire.

1758

5. Fisher, T. ["Comets Not Dangerous"]. **Gentleman's Magazine**, 28 (February 1758), 66-68.

   Letter. The comet is expected to envelop earth in its tail. Disputes notion of comets as ominous, rather they are manifestation of God's power and providence, which He vests in lifeless matter. The heavens declare to us the glory of God.

6. M., N. ["Comet Observed"]. **Gentleman's Magazine**, 28 (June 1758), 253.

   Letter. Dated 20 June 1758. Describes observations of 23 and 27 June. The comet appeared as a small obscure flare, faintly seen through the light of the dawn. A reflecting telescope seemed to magnify it, but at the same time made it more hazy.

7. I., B. ["Comet Observed"]. **Gentleman's Magazine**, 28 (June 1758), 253.

   Letter. Responds to N.M. on same page (see entry 6). If this is the expected return of the comet of 1682 (Halley's) it must have passed perihelion on the 6th. It will, therefore, arrive at its descending node about the 7th July. If it is the comet of 1682 it should have a conspicuous tail--none, however, was observed by the author.

8. G., J. ["Poem"]. **Gentleman's Magazine**, 28 (October 1758), 487.

   Poem.

9. "On the Return of the Expected Comet." **Gentleman's Magazine**, 28 (November 1758), 526-27.

   Presents a synopsis of the comet's history.

10. Bevis, J. "An Account of the Comet Seen in May 1759." **Philosophic Transactions** (Royal Society of London), 51 (1759), 93-94.

> Read 3 May 1759. Comet was observed on 1 May 1759 from 9 to 11 P.M. It was compared by means of "the equatorial instrument." Observed again on 2 May but an increasing moon weakened its light, making the tail and nucleus indistinguishable.

11. Munckley, Nicolas. "An Account of the Same Comet." **Philosophic Transactions** (Royal Society of London), 51 (1759), 94-96.

> Read 3 May 1759. The author first observed the comet on 30 April at Hampstead, fixing its place on 1 May at about 10 P.M. with assistance of a common globe, a quadrant, and Senex's planisphere. The luminous appearance of the comet was evident to the naked eye. It was seen again on 5 and 6 May.

12. Winthrop, John. **Two Lectures on Comets Read in the Chapel of Harvard College ... in April 1759 on Occasion of the Comet Which Appear'd That Month.** Boston: Green and Russell, 1759. 44 pp.

> Expounds the Newtonian theory of comets, asserting that other comets will also periodically return. Holds that the motions of comets are not in straight lines, and their tails are not produced by transmission of light. Morally, comets are viewed to serve to remind men of the power of the "Supreme Governor of the Universe." Published by the "general desire of the hearers" of the lectures.

13. ["Time of the Expected Comet's Perihelion Calculated"]. **Gentleman's Magazine,** 29 (February 1759), 52-54.

> Predicts that comet will pass perihelion about mid-April 1759. Reviews Halley's calculations, and remarks that computing the comet's passage cannot be an exact science.

14. ["On the Present Comet"]. **New American Magazine,** 16 (April 1759), 438.

> Poem. A note refers to the comet, observed April 3rd at about 3 o'clock with naked eye. The tail was estimated to be about 12 or 13 degrees in length and silver in color.

15. Astrophilius. ["Problem Proposed to the

Astronomers"]. **Gentleman's Magazine,** 29 (April 1759), 154.

Letter. Newspapers inform us that the comet has been seen this month in Germany, Paris, and England. It was observed in Paris on the 1st and 2nd near tail of Capricorn. It is easy to figure the comet's perihelion, node, or orbit--but can its nearest point to the earth be calculated?

16. **Boston Evening Post,** 9 April 1759, 3.

For several mornings past a comet has appeared in the northeast and inclines pretty fast to the south. It rises after 3 o'clock, its tail appearing to be about 10 feet long and standing right up. It is supposed to be the same comet predicted for earlier this year.

17. Winthrop, J. **Pennsylvania Gazette,** 1759-04-19, [p. ?].

Letter. Dated 7 April 1759. "The comet, which for some time has been a pretty general topic of conversation; and the prediction of which has excited curiosity in many, and terror in more, has at length ... made its appearance." The author goes on to give the comet's position.

18. "On the Appearance of the Comet in America." **Gentleman's Magazine,** 29 (May 1759), 229.

Poem. An appended note states that the comet was first observed in America at New York on 3 April 1759 in the sign of Pisces. It was silver in color.

19. ["Some New Observation on Cometical Astronomy"]. **Gentleman's Magazine,** 29 (May 1759), 205-09.

Letter. The period of this comet is one and a half years longer than its last because of the influence of Jupiter upon it during its last descent to the sun. The author presents his personal theory of comets.

20. Astrophilius. ["Further Observations on the Present Comet." **Gentleman's Magazine,** 29 (May 1759), 204.

Letter. The path of the comet is traced under the constellation of the Cup, across the Hydra, and up to the Sextant, where it was last night (26 May 1759). It will probably be found a bit northward of its present location in about three weeks.

21. W., J. ["The Comet"]. **Boston Evening Post,** (2 May 1759), 3?.

The comet which appeared in the beginning of April has passed its descending node and continued its route southwestward as expected. It was observed on 22 and 25 April—gives locations in sky. Calculates that comet is better observed in the Southern Hemisphere.

22. Me. **New American Magazine**, (June 1759), 517.

Letter. Dated 20 May 1759, Philadelphia. Gives observations of the comet for 1, 3, 4, 5, and 16 May 1759.

23. Philomathes. **New American Magazine**, (June 1759), 517.

Letter. Addressed to Mr. Sylvanus Americanus (editor). The author first saw the comet on 4 April at 4:00 A.M., located in nearly a direct line with two stars in the right and left arms of the constellation Aquarius.

1831

24. Lubbock, Mr. "Two Notes by Mr. Lubbock, On the Comet Halley." **Monthly Notices of the Royal Astronomical Society**, 2 (14 January 1831), 5-6.

In the first note, Lubbock deduces the elements of orbit for the comet. In the second note he analyzes a paper on the same subject by Rosenberger appearing in **Astronomische Nachrichten**, nos. 180 and 181.

25. Lubbock, J.W. "Note on the Orbit of Halley's Comet." **Memoirs of the Royal Astronomical Society**, 4, Part 2 (1831), 509-16.

Corrects his previous calculations of the comet's elements of orbit (see entry 24). He has since found that the change in the major axis due to the action of the planets Jupiter and Saturn was so great as to affect the values of those elements for 1759.

1834

26. "On the Reappearance of Halley's Comet." **Philosophical Magazine**, 5 (October 1834), 284.

Presents ephemeris for comet for 16 December 1834 till 5 April 1835. Mentions that Rumker thinks that it may be possible to see comet with a telescope as early as December 1834.

27. **Niles' Weekly Register**, 47 (22 November 1834), 185.

Quotes the **Mobile Advocate** as saying that there is

confusion between the comet now coursing through the firmament and Halley's Comet which is expected to be punctual in its return twelve months from now, in November 1835.

28. Olbers, Dr. "Translation of a Paper by Dr. Olbers in the **Astronomische Nachrichten, No. 268,** on the Approaching Return of Halley's Comet." **Monthly Notices of the Royal Astronomical Society,** 3 (December 1834), 68-71.

> Translated by Mr. Galloway. Olbers states that the date of perihelion passage cannot be calculated with certainty. Presents computations using both the 1st and the 15th of November as date of perihelion. It is not thought that the comet will be as brilliant as that of 1811. See also entry 37.

1835

29. "Halley's Comet." **Family Magazine,** 3 (1835), 157-58.

> The comet will be nearest to earth on about 5 or 6 October and in perihelion about 4 November. Recounts the story (apocryphal) of Pope Callixtus III's excommunication of the comet. Speculates as to what would happen should the comet strike the earth--speculates that the earth's rotation would be reversed.

30. "On the Comet of Halley, Sometimes Called the Comet of 1759." **Edinburgh Journal of Natural History and of the Physical Sciences,** 1 (24 October 1835), 11.

> Begins with poem, "History of the Comet to A.D. 1456," by Young. Article is drawn chiefly from the work of Arago and Pontecoulant (see entry 34), giving historical background, including the story (apocryphal) of Pope Callixtus II [sic] exorcizing the comet in a Papal bull.

31. "The Comet of Halley." **Edinburgh Journal of Natural Science,** 1 (1835), 24 ff.

> Describes the orbit of Halley's Comet as elongated with a length of about thirty-six times the distance of the sun from the earth, and at its greatest breadth is about ten times the distance of the sun from the earth.

32. ["Halley's Comet"]. **Foreign Quarterly Review,** 15 (1835), 477-81.

> An article reviewing German language books by Moebius, von Littrow, and Fischer. Provides some astronomical history and predictions for the dates of the comet's

appearance and its path.

33. Drew, John. **Two Lectures on Comets.** London: n.p.,
1835.

One of the two lectures relates the astronomical
history of Halley's Comet.

34. Pontecoulant, Louis G. **A History of Halley's Comet:
With an Account of Its Return in 1835.** Tr.   Charles Gold.
London: J.W.   Parker, 1835.   44 pp.

Translation of **Notice sur la Comete de Halley: Et Son
Retour en 1835.**   Paris, 1835.   This is one of the
important early major works devoted to Halley's
Comet--it is often cited.

35. Seares, John. **The Comet.** London: Simpkin, Marshall
and Co., 1835.   112 pp.

In four parts: 1. Concise introduction to astronomy;
2.   Comets in general;   3.   The principle data of
Halley's Comet;   and 4.   History of Halley's Comet.
Illustrated and explained by tables and diagrams.

36. Zornlin, Rosina Marina. **What Is a Comet, Papa?**
London: J. Ridgeway and Sons, 1835.   72 pp.

The wonders of the heavens are displayed and familiarly
explained for the use of children.   The book is a
product of adults communicating to children the
widespread anxiety over the approaching comet.

37. Olbers, Dr. "On the Approaching Return of Halley's
Comet." **Philosophical Magazine,** 6 (January 1835), 45-52.

Translated from German as in **Astronomische Nachrichten,**
no. 268.   The precise day on which Halley's Comet will
make its nearest approach to the sun cannot be
precisely predetermined.   Gives ephemeris, predicts
perihelion passage for 1-5 November 1835.   See also
entry 28.

38. "The Approaching Comet." **Niles' Weekly Register,** 48
(28 March 1835), 64.

Reprinted from the **Falmouth** (England) **Packet.**   "Relying
on the correctness of principle of cometary influences,
we predict summer of 1835 ... intense heat ...
destroy harvest in some parts of world." Also predicts
earthquakes.

39. "Article VII. [book reviews]." **Edinburgh Review,** 61
(April 1835), 82-125.

A Review of Arago's, **Des Cometes en General, et en**

Particulier des Celles qui doivant Paraître en 1832 et 1835, Paris, 1834; G. Pontecoulant's, Notice sur la comete de Halley et son Retour en 1835, Paris, 1835; and Sir J.F.W. Herschel's, Observations on Biela's Comet, London, 1833. "Notwithstanding the discovery of the periodic comets of Encke and Biela, still the comet of Halley maintains a paramount astronomical interest; and may be considered to stand alone in exhibiting those physical phenomena which seem to be the exclusive characteristics of the class to which it belongs."

40. "Halley's Comet." London Times, 23 April 1835, 5.

Column d. Herschel says that Halley's Comet which is positively expected in August of this year, will not be visible because it has long since changed its orbit. A report by English astronomers on this important subject is awaited.

41. "The Comet of 1835." Burlington Free Press [Burlington, Vermont], 15 May 1835, [p. ?].

Notes that the "the comet of Halley" is expected in November. Presents extract from Pontecoulant's book on the comet, giving a synoptic history and astronomical description of the comet.

42. "Halley's Comet." Albion, 3 (13 June 1835), 192.

Abstracted from the April 1835 Edinburgh Review (see entry 39). Describes the low probability of the earth colliding with the comet. Also reviewed are some old comet superstitions.

43. "Halley's Comet." American Magazine, 1 (July 1835), 495.

Nitpicks about the accuracy of Halley's 75 year prediction, pointing out that the period between apparitions of the comet is really closer to 76 years.

44. "Halley's Comet." American Magazine, 1 (August 1835), 535-36.

Discusses variance in predictions of astronomers as to the month of the comet's appearance. This short article purposefully understates the importance of Halley's Comet.

45. "Halley's Comet." London Times, 7 August 1835, 5.

Column d. Quoting from the British Almanack [sic]--provides locations of the comet from 7 August 1835 through 7 February 1836. The best time to look for it with the naked eye will be the first ten days of October 1835.

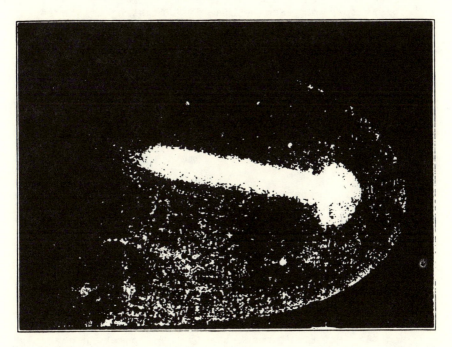

1. Drawing of the comet at its 1835/36 apparition by John Herschel at the Cape of Good Hope. See entries 118, 126, and 130.

46. G., J. "Halley's Comet." **Athenaeum**, No. 407, 15 August 1835, 624.

Letter. "Many absurd statements having appeared in the public journals relating to Halley's Comet ..." the author presents an ephemeris for 15 August through 24 September 1835. It is sent for no other than a popular purpose, the astronomer needing no added assistance.

47. "The Comet." **London Times**, 22 August 1835, 5.

Column c. The comet has been sighted by the observatory at Rome--this first sighting is viewed as an important fact. The light of the comet is described as being very feeble since weather has not been yet sufficiently favorable to permit exact observation.

48. South, James. "Halley's Comet." **London Times**, 24 August 1835, 3.

Column b. Letter. Gives news of his sighting of the comet at his Kensington observatory. Provides the exact position at the time of sighting.

49. Lubbock, J. W. "Halley's Comet." **London Times**, 25 August 1835, 6.

Column f. Letter. Presents note from Dr. Hussey giving location of the comet at the time of his sighting on 23 August. Lubbock comments that Hussey's information is in agreement with the ephemeris published in the 1835 **Nautical Almanac**.

50. Airy, Professor [George]. "Halley's Comet." **London Times**, 31 August 1835, 3.

Column b. Reprinted from the **Cambridge Chronicle**. Gives Pontecoulant's calculation for the perihelion passage, right ascension, and declination, as 7, 10, and 16 November respectively. The comet is now only visible with a telescope of at least 6 inches.

51. "Apostrophe to the Approaching Comet." **New Monthly Magazine**, 44 (September 1835), 89-90.

Poem.

52. South, J. "To the Editor of the Nautical Magazine." **Nautical Magazine**, 4 (September 1835), 565.

Letter. Having found Halley's Comet yesterday morning (23 August), South forwards results of his observations for 22 and 23 August. The comet is extremely faint, nearly round, and perhaps two minutes in diameter.

53. South, James. "Halley's Comet." **Nautical Magazine**, 4

(September 1835), 564-65.

> Reprinted from the **Companion to the British Almanac**. Presents Halley's calculations for the comets of 1531, 1607, and 1682 in tabular format. Notes that a diminution of light has been noted in other periodic comets. It must now be seen if the brightness of Halley's Comet will diminish with each apparition.

54. **London Times**, 1 September 1835, 6.

> Column e. Reprinted from **Mechanics Magazine**. The nearest approach of Halley's Comet will be between 7 and 8 October 1835 and it will reach perihelion on 17 November. Recounts previous apparitions, incorrectly giving dates as 1005 and 1080--actually 1066.

55. Cooper, Edward J. "To the Editor of the Standard." **London Times**, 1 September 1835, 6.

> Column e. Letter. Recounts observation of the comet on 26 August using a telescope of 13" aperture. The nucleus was perfectly distinguishable.

56. South, James. "Halley's Comet." **London Times**, 1 September 1835, 6.

> Column e. Letter. That the comet is without tail should not be construed to indicate that it has undergone a physical change since it last visited. One does not need a large telescope for viewing--42-inch achromatic with 2 3/4-inch aperture is sufficient or 28 inch with 2-inch aperture hand held.

57. Thompson, C. "Halley's Comet." **London Times**, 5 September 1835, 4.

> Column d. Letter. The author is assistant astronomer at Dublin Observatory. Observed comet at 2:00 A.M. on morning of 30 August. Offers his observation as verification of Pontecoulant, Damoiseau, and Lubbock. The letter is addressed to the editor of **Saunders's Newsletter**.

58. "Halley's Comet." **London Times**, 11 September 1835, 5.

> Column b. Reprinted from the **Hampshire Chronicle**. Gosport Observatory reports that the comet rises in the northeast at 10:45 P.M. It is an old comet--in existence upwards of 500 years. Our admiration of the present visit is excited by its curious historical associations.

59. **Niles' Weekly Register**, 49 (12 September 1835), 20.

> Professors Olmsted and Loomis of Yale College claim the

honor of first having discovered Halley's Comet on 31 August 1835--based on a card that they published in the **New Haven Herald** (quoted in its entirety).

60. **Niles' Weekly Register,** 49 (12 September 1835), 20.

Quotes from a letter from a Mr. Rogers which originally had appeared in the **Bristol** (England) **Gazette,** 13 June 1835. "The comet Halley now arrests the attention of the astronomical world." Perihelion is predicted for about 4 November 1835.

61. Airy, [George]. "Halley's Comet." **Penny Magazine,** 4 (12 September 1835), 360.

Reprint of an account communicated to the **Cambridge** (England) **Chronicle.** As early as 6 August the comet was seen at Rome. The nearness of its actual appearance to the predicted time of appearance proves that there is no unknown planet of great bulk.

62. Redding, Cyrus. "Halley's Comet." **Athenaeum,** No. 411, (12 September 1835), 699.

Letter. Suggests that the comets mentioned in the years 842 and 1222 A.D. were also Halley's Comet. The author plots other probable apparitions of Halley's Comet at 76 year intervals.

63. Plana. "Halley's Comet." **London Times,** 18 September 1835, 3.

Column d. Letter. Reprinted from **Piedmont** (Italy) **Gazette;** the author is Royal Astronomer at the Observatory of Turin, Italy. The comet is seen for first time on 1 September--gives measurements. On 1 September comet was calculated to be 48 million French leagues from the earth.

64. E. "Halley's Comet." **London Times,** 26 September 1835, 7.

Column b. Letter. The writer expresses confusion over the apparent inability of astronomers to agree on a single date for comet's perihelion. He calculates the speed of the comet to be 2000 miles per second, based on published information. He asks if this is correct.

65. South, J. "Halley's Comet." **London Times,** 30 September 1835, 4.

Column f. The comet was first seen by the author at Kensington with the naked eye on 25 September at 10:30 A.M. He gives the location of the comet and reference points for use by other observers. The comet was preceded by a small star, probably "occulted" in the comet.

66. "Halley's Comet." **American Journal of Science,** 29 (October 1835), 155-56.

States that the comet was first observed in 1835 by Olmsted and Loomis at Yale College on 31 August 1835. Predicts perihelion to be on 16 Nov 1835.

67. South, J. "Mr. Jump's Account of Halley's Comet." **London Times,** 2 October 1835, 3.

Column b. Letter. Presents an ephemeris of Halley's Comet based on a perihelion of 16 November 1835. "With many thanks to you for the promptness with which you have given my communications that publicity which I must in vain sought elsewhere."[sic]

68. "History of Halley's Comet." **Poorman's Guardian,** (3 October 1835), 693-94.

A rendition of the "malignant effect that in days of ignorance and superstition were believed to have marked its onward course, or its departure from our firmament."

69. Common Sense. "To the Editor of the Times." **London Times,** 5 October 1835, 7.

Column b. Letter. Asks that the **Times's** correspondents favor readers with a more popular, less technical, account of the comet's path. The author asserts that 90% of the **Times's** readers would find this very acceptable.

70. Nauticus. "Halley's Comet." **London Times,** 5 October 1835, 7.

Column b. Letter. In reference to Sir James South's communication in the **Times** of 2 October (see entry 67). The author questions the authority of the compilations upon which South's ephemeris is based.

71. "Comet." **New Jersey Journal** [Elizabeth, New Jersey], 6 October 1835, [p. ?].

The comet will rise on 1 October at about 9:15 P.M. From the 7th to the 13th of the month it will be within a circle of perpetual apparition, revolving around the pole without descending below the horizon. At end of month it will appear only in evening.

72. An Old Correspondent. "To the Editor of the Times." **London Times,** 6 October 1835, 3.

Column c. Letter. Having witnessed the anxiety of the

public to see Halley's Comet, the author offers to place his achromatic telescope of 12-feet focus and 8-inch aperture in any convenient situation for public use. He asks under what restriction admission to use the telescope should be allowed.

73. South, J. "Halley's Comet." **London Times**, 6 October 1835, 3.

Column c. Letter. In response to Nauticus in the **Times** of 5 October (see entry 70). South gives bearings for every hour of the next 4 days for the comet observed from the latitude of London. The comet will be visible to the naked eye, but it will be unlikely to be found unless one knows where to look for it

74. Gamma Draconis. "To the Editor of the Times." **London Times**, 7 October 1835, 1.

Column f. Letter. Comments on the "very liberal offer" of an Old Correspondent in the **Times** of 6 October (see entry 72). Suggests that a committee of the Royal Astronomical Society be appointed to take charge of the telescope. Only members of chartered scientific societies should be allowed to use the telescope.

75. South, J. "Halley's Comet." **London Times**, 8 October 1835, 5.

Column c. Letter. Presents a chart of locations of the comet for 10, 11, 12, and 13 October 1835.

76. **The Bee** [New Orleans, Louisiana], 9 October 1835, 2.

Column 2. A two line blurb: "The comet is now visible between one and two o'clock in the morning, a little north of the east cardinal point." The mention of the comet at all is significant, since **The Bee** devoted its attention at this time almost exclusively to matters of commercial interest.

77. Sloane, James. "Halley's Comet." **London Times**, 9 October 1835, 3.

Column d. Letter. The comet is of a dull gray lead color without any distinct nucleus, a bit bearded, probably from its tail (if it has any) being turned towards the earth. It can be seen but not distinguished by the naked eye.

78. Stratford, W.S. "To the Editor of the Times." **London Times**, 13 October 1835, 3.

Column c. Letter. In response to Sir James South's

correspondence in the **Times** of 2 October and Nauticus's on 5 October (see entries 67 and 70 respectively). The author is Superintendent of the **Nautical Almanac**; he takes umbrage at the criticisms directed at the **Almanac** and offers rebuttal.

79. **Sentinel of Freedom** [Newark, New Jersey], 16 October 1835, 1.

Halley's Comet could be seen distinctly with the naked eye during the last week. The comet which spread terror among the people of the 15th century can be traced back to the commencement of the Christian era.

80. South, J. "Halley's Comet." **London Times**, 16 October 1835, 5.

Column c. Letter. To answer the inquiries of many of your readers, Halley's Comet during hazy weather or through clouds cannot be well missed by anyone looking for it in the place indicated by its ephemeris. Author has seen the nucleus--it looks like a little planetary disk.

81. "The Comet." **Christian Messenger**, 14 (17 October 1835), 247.

Short announcement. The comet is now clearly visible near the northern part of its course, in the constellation of the Great Bear. It is about 5 days behind prediction. This is not a great mistake in reckoning considering a period of more than 76 years.

82. ["Halley's Comet"]. **The Friend**, 9 (17 October 1835, 15.

Halley's Comet is now to be seen near the right foot of Hercules in a straight line with Beta Alioth and Mizar in the Great Bear. This is the comet's second return since Halley conjectured its periodicity. It may be safely predicted to return again in 1910-11.

83. Littrow, C.L. **London Times**, 17 October 1835, 3.

Column b. The comet is visible to the naked eye as a star of the third magnitude with scarcely a nebula about it. Its largest diameter is about half that of the sun, with a bright, pretty large, scintillating nucleus, but still without any considerable tail.

84. South, J. "Halley's Comet." **London Times**, 19 October 1835, 3.

Column c. Letter. Includes Rosenberger's latest ephemeris of the comet, calculated at the latitude of Berlin for October 1835. South remarks that the increase in the comet's brilliance is important because it may now be observed under unfavorable weather conditions.

85. South, J. "Halley's Comet." **London Times**, 21 October 1835, 5.

> Column d. Letter. Presents a copy of "Rosenberger's Former Ephemeris of Halley's Comet," which he says is unavailable at any price in England. The comet was visible at Kensington where its tail subtended at an angle of 38 degrees.

86. [South, James]. "Halley's Comet." **London Times**, 22 October 1835, 3.

> Column f. The comet has two tails, one is in line with the other; they lie in the south preceding and the north following quadrants, and are inclined parallel, at an angle of 75 degrees. The nucleus is very bright.

87. "Sir James South." **London Times**, 23 October 1835, 3.

> Column b. Sir James South invites fifty boys and their masters from Christ's Hospital to his home for dinner and to view the comet through his telescope.

88. "The Comet." **Workingman's Advocate**, 7 (24 October 1835), [3].

> Short announcement. Halley's Comet is now plainly visible in the northwest during the first part of the night. It has an extra tail. Quotes **Schenectady** (New York) **Reflector** as asserting that this is the first time the comet has been observed with more than one tail.

89. "To the Celestial Stranger--The Comet." **Cincinnati Advertiser and Ohio Phoenix**, 24 October 1835, 2.

> Poem.

90. Rosenberger, A. "Halley's Comet." **London Times**, 26 October 1835, 5.

> Column d. Ephemeris for 20 October through 22 November 1835. Forwarded to the **Times** by James South having received it from H.C. Schumacher, Astronomer Royal of Denmark.

91. "Halley's Comet." **Delaware State Journal** [Wilmington, Delaware], 27 October 1835, [p. ?].

> A copy of a record kept by a lady of Wilmington, Delaware. Observer's notes for 13-22 October 1835. Observations were on each night taken at between 7 and

8 P.M. Declination and right ascensions are given plus positions in relation to constellations.

92. "The Comet." **London Times**, 28 October 1835, 2.

Column e. A capsule from a Paris newspaper. Reports Arago's observations relating to the physical makeup of the comet.

93. South, J. "Halley's Comet." **London Times**, 29 October 1835, 3.

Column d. Letter. Reports that he, too, on several occasions has seen a small fan of light independent of the comet's large ordinary tail, just as Arago has observed.

94. "Halley's Comet." **Atkinson's Casket**, (November 1835), 646-47.

Extracts reprinted from April 1835 **Edinburgh Review** (see entry 39). Gives brief astronomical introduction to comet and scheduled appearance. Discussion of formation of comets and probability of a collision with the earth. Mention is made of some of the "absurd" effects attributed to comets.

95. **Delaware Gazette and Watchman** [Wilmington, Delaware], 3 November 1835, 1.

Astronomers cannot agree on whether the comet has one or two tails. Quotes **Boston Post**: "truth is, there is only one, but that is so long the Comet is obliged to double it to prevent its dragging upon the earth, and people from catching hold of it."

96. "The Comet." **Christian Intelligencer**, 15 (6 November 1835), 166.

The comet about which so much has been said, has appeared and disappeared, and this globe still continues to roll upon its axis. The credulous and superstitious have made themselves miserable—prophets who predicted calamity proved themselves to be ignoramuses.

97. South, J. "Halley's Comet." **London Times**, 10 November 1835, 3.

Column b. Presents ephemeris for November 1835, computed at Greenwich Mean midnight—from the office of the Nautical Almanac.

98. Pearson, D. "Extract of Letter from Dr. Pearson to Francis Baily, Esq., Containing an Observation of Halley's Comet." **Monthly Notices of the Royal Astronomical Society,**

3 (13 November 1835), 130-31.

> Observations made on 19th, 28th, 29th, and 30th October 1835 are reported. Pearson explains his methodology and computations.

99. T. "Halley's Comet." **London Times**, 18 November 1835, 3.

> Column f. Letter. The correspondent asks Lt. Stratford, R.N., Superintendent of the **Nautical Almanac,** what three controlling observations were used by the calculator of that office's ephemeris for November 1835, because it differs from other ephemerides.

100. Stratford, W.S. "Halley's Comet." **London Times,** 20 November 1835, 3.

> Column e. Letter. The Superintentent of the **Nautical Almanac** takes umbrage at T's letter (see entry 99). Stratford objects to T's anonymity, suggesting direct queries would be answered more quickly--that is, if knowledge is the goal.

101. Taylor, J. "Halley's Comet." **London Times,** 24 November 1835, 5.

> Column c. Letter. T (see entry 99) reveals himself to Lt. Stratford (see entry 100). Taylor views Stratford as discourteous to reasonable inquiries. He still wants to know what the three controlling observations for the November ephemeris are.

102. Stratford, W.S. "Halley's Comet." **London Times,** 27 November 1835, 4.

> Column a. Letter. Replies to Taylor (see entries 99 and 101). Stratford provides the elements of the calculations for the **Nautical Almanac's** November ephemeris for Halley's Comet.

103. South, J. "Halley's Comet." **London Times,** 30 November 1835, 5.

> Column e. Letter. Supplies A. Rosenberger's latest ephemeris for 21 November 1835 through 12 January 1836. South has received ephemeris from H. C. Schumacher, Astronomer Royal of Denmark.

104. Stratford, W.S. "To the Editor of the Times." **London Times,** 30 November 1835, 5.

> Column e. Letter. Having received Professor Schumacher's circular of 20 November 1835 containing Professor Rosenberger's amended elements for Halley's

Comet for November 1835 through 12 January 1836 (see also entry 103), Stratford is now able to present the results of comparison orbits, which he gives.

105. "Article VII. [book reviews]." **Quarterly Review**, 55 (December 1835), 195-233.

Review of von Littrow's, **Ueber den Halleyschen Kometen**, Vienna, 1835; and von Encke's, **Ueber den Halleyschen Kometen**, Berliner Jahrbuch, 1835. Provides an historical and astronomical overview of Halley's Comet while discussing the virtues of the two books.

106. Draper, E. "Halley's Comet--1760." **Southern Literary Messenger**, 2 (December 1835), 9-10.

Poem. The anachronism is obvious.

107. Taylor, John. "Halley's Comet." **London Times**, 3 December 1835, 7.

Column c. Letter. Explains his objection to the **Nautical Almanac**'s method of calculating the ephemeris for Halley's Comet (see entries 99 and 100). He tangentially mentions that the comet appeared at the time of the birth of Alexander the Great and Mithradates.

108. Morrison, Richard James. **Observations on Dr. Halley's Great Comet Which Will Appear in 1835**. London, 1835.

A history of phenomena attending the return of Halley's Comet for six hundred years past. Predicts calamities that will happen in conjunction with the 1835 apparition. See also entry 164.

1836

109. Smyth, W.H. "Observations of Halley's Comet." **Memoirs of the Royal Astronomical Society**, 9 (1836), 229-46.

Observed at Bedford, England, 24 August through 21 October 1835. "The return of Halley's Comet is a phenomenon of universal interest." The comet was observed using annular micrometer. Smyth explains his methodology and gives determinations of the comet's geocentric places. Read 10 June 1836.

110. "Arago on Comets." **North American Review**, 42 (January 1836), 196-216.

Reviews two books by Arago in light of the approach of

Halley's Comet. In addition to evaluating the books at hand, a bit of the comet's history is given, including the story (apocryphal) of Pope Callixtus' "cursing" the comet in 1456. There is a catalog of some of the disasters attributed to cometary influences.

111. "Observations for Determining the Positions of Halley's Comet." **Monthly Notices of the Royal Astronomical Society,** 3 (8 January 1836), 142.

Four sets of sextant observations taken at sea by Owen, Cannon, Dursterville, and Lawrence.

112. Boguslawski. ["Re: Letter on Observance of ..."]. **Monthly Notices of the Royal Astronomical Society,** 3 (8 January 1836), 142.

Letter dated 28 Aug 1835, Royal Observatory, Breslau. The comet was observed on 21, 22, 24, 25, 26 Aug 1835.

113. "The Comet." **London Times,** 20 January 1836, 1.

Column f. A short blurb reporting that M. J. Mueller of the observatory at Geneva, Switzerland again saw Halley's Comet on the night of 31 December. It was very faint indeed, but was precisely in accordance with calculations.

114. "Halley's Comet." **London Times,** 23 January 1836, 7.

Column c. Dateline Augsburg, 14 January. The comet, after having been invisible for a month, was once again seen from the observatory at Milan on 30 December. It will be visible until April. The earth will approach the comet again in March but at a greater distance.

115. Gorham, George Cornelius. "Halley's Comet." **London Times,** 26 January 1836, 3.

Columns a-b. Letter. Halley's Comet is the only heavenly body of known period, which makes its appearance but once, or at most twice, within the terms of a human life; its visits, like angels, are few and far between. For each recorded apparition a brief discussion of concurrent historical circumstances is given.

116. Prince Edward. "Halley's Comet." **Southern Literary Messenger,** 2 (March 1836), 235.

Poem.

117. "Halley's Comet." **London Times,** 11 March 1836, 5.

Column c. Dateline Buenos Aires, 14 December. Various captains of vessels lately arrived here report that

Halley's Comet has been visible in this hemisphere for the past 6 weeks.

118. Herschel, Sir John. ["Re: Letter on Observation of ..."] **Monthly Notices of the Royal Astronomical Society,** 3 (8 April 1836), 190.

Herschel reports in letter dated 3 February 1836 that the comet keeps him awake all night, every night ... that it is the most beautiful thing he ever saw in a telescope. The most surprising thing about the comet is the enormous increase in the comet's dimensions.

119. "Comets and Eclipses." **Evangelical Magazine and Gospel Advocate,** 7 (14 May 1836), 160.

Halley's achievement in predicting the return of the comet represents a triumph of intellect that will govern our moral destiny. Such an approach will banish superstition and false belief. Reprinted from **The Knickerbocker.**

120. Smyth, W.H. "Observations of Halley's Comet, Made at Bedford." **Monthly Notices of the Royal Astronomical Society,** 3 (10 June 1836), 198-99.

Presented are 50 places of the comet on 18 different days from 24 August to 21 October 1835. With the exception of 3 meridian observations and four with a double-wire micrometer, the observations were made with an annular micrometer. See also entry 109 for fuller rendition.

121. Loomis, Elias. "Observations on the Comet Halley, Made at Yale College." **American Journal of Science,** 30 (July 1836), 209-21.

The 1835 apparition was first observed in Rome in August 1835. Gives an account of various other sightings in chronological order. The information offered here says less about the actual sightings than it does about the limitations of communication at the time. In 1759 the comet attracted little attention.

122. "Effects of Halley's Comet." **London Times,** 28 October 1836, 3.

Column c. From the **Caledonian Mercury.** Infers a connection between the comet and climatic changes: 1280 had great heat; 1305 saw the Baltic Sea freeze for 14 weeks; 1455 had great rains; in 1531 floods hit Rome, Antwerp, and Lisbon; and 1835 was cold and wet.

123. Maclear, Mr. [Thomas]. ["Extract of Letter Re: Original Circle and Transit Observations of Halley's Comet

Since January."] **Monthly Notices of** the Royal Astronomical **Society,** 4 (11 November 1836), 3.

The number of meridian observations thus obtained is upwards of thirty. The reductions will be forwarded in a short time.

1837

124. Joslin, B.F. "Observations on the Tails of Halley's Comet, As They Appeared at Union College, Schenectady, N.Y." **American Journal of Science,** 31 (January 1837), 142-55.

The head and length of the tail has been perceived to be less bright on each successive return of the comet. Discusses factors in determining the brightness of the tail. Observations of 4 through 25 October 1835 are presented.

125. Joslin, B.F. "Additional Remarks on the Tails of Halley's Comet." **American Journal of Science,** 31 (January 1837), 324-32.

About the brush of light observed 12 October 1835 at Union College, Schenectady, NY. Also presents a summary of other published articles relating to the tail of Halley's Comet.

126. Herschel, Sir John. "Observations of Halley's Comet, After Its Perihelion Passage." **Monthly Notices of the Royal Astronomical Society,** 4 (13 January 1837), 25.

Abstract of paper (see entry 130) containing observations of Halley's Comet, compared with certain stars near it, from 25 January to 5 May 1836, made in the southern hemisphere--describes the striking increase in comet's size after the perihelion passage.

127. Maclear, Mr. "Observations on Halley's Comet." **Monthly Notices of the Royal Astronomical Society,** 4 (14 April 1837), 73-74.

Maclear is astronomer at Cape of Good Hope observatory. The comet was observed from 16 February through 5 May 1836. The faintness of the comet compelled Maclear to various expedients which are explained. The British government ordered this paper published at public expense.

128. Taylor, T.G. "Observations of Halley's Comet, Made at the Observatory at Madras." **Monthly Notices of the Royal Astronomical Society,** 4 (12 May 1837), 79-80.

Fourteen observations were made from 19 February through 21 March 1836. The comet was extremely faint and therefore prevented more accurate results.

1838

129. Maclear, Mr. "Observations of Halley's Comet, Made at the Royal Observatory, Cape of Good Hope, in the Years 1835 and 1836." **Memoirs of the Royal Astronomical Society,** 10 (1838), 91-155.

Read 14 April 1837. One of the most complete of published observations of the 1835-36 apparition. Replete with extensive narrative and tabular data.

130. Herschel, Sir J.F.W. "Observations of the Comet of Halley, After the Perihelion Passage in 1836: Made at Feldhausen, Cape of Good Hope." **Memoirs of the Royal Astronomical Society,** 10 (1838), 325-35.

Read 13 January 1837; letter dated 18 October 1836. Herschel lost sight of the comet after 5 May 1836. Presents a list of stars compared with Halley's Comet and notes from his journals; also right ascension and declination at time of opposition in 1836.

131. Taylor, T.G. "Right Ascension and Declination of Halley's Comet Near the Time of Opposition in 1836." **Memoirs of the Royal Astronomical Society,** 10 (1838), 335.

Calculations of right ascension and declination taken at Madras, India. The right ascensions are accurate to one second of time, and the declinations to about 20 or 30 seconds of space; the faintness of the comet prevents a more accurate observation.

1842

132. Mueller, [M.J.]. "Observation of Halley's Comet, Made at the Observatory in Geneva in the Years 1835 and 1836." **Monthly Notices of the Royal Astronomical Society,** 5 (14 January 1842), 129-30.

Observations made on fifty-two nights, 31 August 1835 through 7 May 1836--31 before perihelion and 21 after perihelion at the observatory of Geneva, Switzerland. Because of observatory's height above sea level the comet was visible longer than from most of Europe.

1846

133. Astronomer Royal [George B. Airy]. "Reduction of
the Observations of Halley's Comet, Made at the Cambridge
Observatory in the Years 1835 and 1836." **Monthly Notices of
the Royal Astronomical Society,** 7 (11 December 1846),
188-89.

Observations made chiefly with a 5-foot equatorial
telescope. Lists nine bases upon which the reduction
has been accomplished. See also entry 134 for fuller
version.

1847

134. Airy, G.B. "Reductions of the Observations of
Halley's Comet Made at the Cambridge Observatory in Years
1835 and 1836." **Memoirs of the Royal Astronomical Society,**
16 (1847), 337-77.

Complete and meticulous tabular data for the
observatory's observation of the comet during the
1835-36 apparition. Read 8 January 1847.

1850

135. Hind, J.R. "On the Past History of the Comet of
Halley." **Monthly Notices of the Royal Astronomical Society,**
10 (11 January 1850), 51-58.

Reviews astronomical and historical scholarship devoted
to the comet. Using Chinese records various Halley
apparitions are verified. Hind's perspective is more
astronomical than historical. This is a most important
piece of early Halley's Comet scholarship.

1852

136. Hind, J. Russell. **The Comets: A Descriptive
Treatise Upon Those Bodies.** ... London: Parker and Son,
1852. pp. 35-57.

Chapter 4 is devoted to the astronomical history of
Halley's Comet. Hind expands on the scholarship
reported in the pages of the **Monthly Notices of the
Royal Astronomical Society** (see entry 135).

HALLEY'S COMET IN 1066, ON THE BAYEUX TAPESTRY.

2. Comet panel from the Bayeux tapestry is a current theme and illustration in the Halley's Comet literature. See entries 321, 377, 389, 548, 1036, and 1068.

HALLEY'S COMET OF 1465.

3. Drawing of Halley's Comet in "1465" (1456) from *Belford's Annual*. Note the error in date and anachronism in clothing. See entry 144.

1861

137. Bronte, Rev. Patrick. "On Halley's Comet, In 1835."
Bradfordian, No. 2 (1 August 1861), 176.

   Poem. [Bradford, Yorkshire, England].   Reprinted in
   Popular Astronomy, 12 (October 1904), p. 571.

1871

138. Williams, John. **Observations of Comets From B.C.**
611 to A.D.   1640:   Extracted From the Chinese Annals.
London: For the Author, 1871. 169 pp.

   A chronological listing of comet observations from
   various   Chinese   sources.   Dates   and   locations
   corresponding to apparitions of Halley's Comet provide
   reaffirmation   of   time, and in some cases, position of
   the comet. This is an important early piece of comet
   scholarship that will be used in future efforts to
   corroborate the times of apparition of Halley's Comet.

1875

139. "The Next Return of Halley's Comet." **Nature,** 11   (11
February 1875), 286-87.

   A short review of Pontecoulant's prediction (1864) of
   the return of Halley's Comet for 1910.

1877

140. Tennyson, Alfred. **Harold: A Drama.**   London: H.S.
King, 1877. I,i,1-53.

   See Act I, scene i, lines 1-53.   "Yon grimly-glaring,
   treble-brandish'd scourge of England! ... Three rods
   of blood-fire up yonder mean/The doom of England and
   the   wrath of Heaven?" An intentional anachronism has
   the comet appearing before the death of King Edward the
   Confessor.   Tennyson uses   the   dread associated with
   comet lore correlatively to set the play's tone at the
   outset.

1886

141. Denning, W.F. "Meteor Shower of Halley's Comet."
Monthly Notices of the Royal Astronomical Society, 46 (14
May 1886), 396-98.

Meteor showers sighted 20 April-6 May 1886 have a
radiant agreeing so closely with the radiant of
Halley's Comet, and is of such pronounced and definite
character, that the identity of the two orbits seem
placed beyond a doubt.

142. Lynn, W.T. "Cursing a Comet At Constantinople."
Notes and Queries, 1 (12 June 1886), 471-72.

When the oft repeated story of Pope Callixtus III
exorcising a comet was invented is difficult to say.
Chroniclers speak of two comets in 1456, but it is
probably Halley's before and after perihelion. Lynn
discusses the special prayers for divine assistance
that were offered in face of the comet.

143. Lynn, W.T. "Halley's Comet At Its Appearance In
1759." **Observatory,** 9 (August 1886), 284-86.

Lynn takes issue with the widely held notion that
Palitzsch, when first sighting Halley's Comet on
Christmas day 1759, did so with the naked eye. He
points out that evidence indicates that Palitzsch used
a telescope of eight focal feet.

1887

144. "Halley's Comet." **Belford's Annual 1887-8.** Chicago:
Belford and Clarke, 1887.. pp. 170-71.

A short paragraph describing how "the appearance of
this particular comet in the year 1465 [sic], just as
the Turks had become masters of Constantinople, was
regarded with superstition and dread. ..." People in
the accompanying illustration wear nineteenth-century
style clothing.

1891

145. Lynn, W.T. "Sir William Lower's Observations of
Halley's Comet in 1607." **Observatory,** 14 (October 1891),
347-48.

Letter. Clarifies misinformation regarding Lower's
observations made while crossing over the Bristol

Channel to Kidwelly. Observations were made with a land surveyor's cross-staff from 22 September to 6 October 1607.

1894

146. "1910 Return of Halley's Comet." **Nature,** 49 (8 March 1894), 442.

Reports that "Prof. Glasenapp announces that the computing bureau established by the Russian Astronomical Society has undertaken the calculation of the true path of Halley's comet, with a view to predicting the exact date of the next return. He hopes that astronomers acquainted with published observations of the comet will communicate the information to the society."

1897

147. Ravene, Gustave. "The Appearance of Halley's Comet in A.D. 141." **Observatory,** 20 (May 1897), 203-05.

Confirms Hind's assertion (see entry 135) that the comet of 141 A.D. was, in fact, Halley's Comet. Recounts Chinese observations of the 141 apparition, calculating the perihelion to be anywhere between 23 March and 6 April 141 A.D.

1898

148. [Neward, E.V.] "Famous Comet." **Quarterly Review,** 188 (July 1898), 113-38.

Review of R.A. Proctor's **Old and New Astronomy.** Essay provides some historical associations with comets and "modern" astronomical history of observation Halley's Comet. Mentions that Halley used Flamsteed's observations in calculating his prediction of the comet's periodicity.

149. Lynn, W.T. "The Papal Bull Against a Comet." **Notes and Queries,** 2 (24 December 1898), 517.

States that it is evident that the story of a Papal bull exorcising the comet is an invention of a later age. There were only four bulls issued by Callixtus III and none mentions the comet.

1902

150. Smart, D. "Halley's Comet: 1910 Return." **Journal of the British Astronomical Association**, 12 (January 1902), 134-36.

An approximate eight-day ephemeris for January to the end of August 1910. There is also a diagram projecting the orbit of Halley's Comet and the earth in the plane of the earth's equator.

151. Denning. W.F. "The Meteoric Shower of Halley's Comet." **Journal of the British Astronomical Association**, 12 (February 1902), 175-76.

Emphasizes that little is known about Aquarid meteor showers thought to accompany Halley's Comet. Suggests that further observation is required in order to fix the date of the maximum shower, and the exact place and duration of the radiant. The velocity of the meteors is about 25 miles per second.

152. Denning, W.F. "The Meteoric Shower of Halley's Comet." **Journal of the British Astronomical Association**, 12 (May 1902), 288-89.

Letter. Predicts the probable time for Halleyan meteors to be the night following 17 May. During the first and third weeks of May, in ensuing years, it will be interesting to watch skies for the long-pathed streak-leaving Aquarids.

1903

153. [Crommelin, A.C.D.]. ["From an Oxford Note-book"]. **Observatory**, 26 (March 1903), 338.

Solicits information about work done recently on the orbit of Halley's Comet for a correspondent who is thinking about making an elaborate investigation of the comet's perturbations.

1904

154. McPike, Eugene Fairfield. "Halley's Comet." **Popular Astronomy**, 12 (December 1904), 685.

A note referring to Halley's discovery of the identity of the comet of 1682, thereafter known as Halley's Comet.

155. Izzard, W.H.   "Halley's Comet." Journal of the
British Astronomical Association, 15 (February 1905), 168.

   Letter. Izzard asks for clarification between two
   accounts of Palitzsch's sighting of the comet on
   Christmas day 1758. Was it with the naked eye or with
   an eight-foot telescope?

156. Observatory, 28 (March 1905), 141.

   The announcement of a prize of 1000 Marks, offered by
   Germany's Astronomische Gesellschaft, "for the best
   determination of the positions of Halley's Comet for
   the year of its return." See entry 297 for winning
   essay.

157. Denning, W.F. "George Palitzsch and Halley's Comet."
Journal of the British Astronomical Association, 15 (March
1905), 203-04.

   Responds to query in earlier issue (see entry 155)
   asking whether Palitzsch discovered comet in 1758 with
   or without aid of telescope. Reaffirms that it was
   with telescope. Traces the source of the naked eye
   misinformation to Delambre.

158. McPike, Eugene F.   "A Bibliography of Halley's
Comet." Observatory, 28 (March 1905), 141.

   Letter.   Solicits notices for manuscript titled
   "Halley's Comet: Its Past History and 1910 Return," a
   bibliography to be published in 1905.

159. McPike, Eugene Fairfield. "Halley's Comet and Its
Discoverer." Observatory, 28 (June 1905), 256-57.

   There is a reference to a vision of a comet in
   Josephus's Jewish Wars (66-85) and in the Talmud there
   is mentioned a comet (Horijoth 10a), where Rabbi Joshua
   says to Rabbi Gamaliel, "Once in 70 years there appears
   a star which is misleading the ships (the sailors).
   ..." Mc Pike finds it remarkable to find the existence
   at so early a date documentary evidence of the idea
   that a comet possesses orbital motion and reappears at
   stated intervals. See also entry 529.

160. McPike, Eugene Fairfield. "Halley's Comet, It's Past
History and 1910 Return." Smithsonian Institution
Publication, no. 1580 (10 June 1905), 69-74.

   Reprinted in Smithsonian Miscellaneous Collections,
   vol. 48 (1905), pp. 69-74. A brief historical
   overview of the astronomical history of the comet as

well as some interesting historical events associated by the public mind with its appearances.

1906

161. Lynn, W.T. "Perihelion Distance of Halley's Comet." **Observatory**, 29 (January 1906), 67-68.

Corrects previous errors in calculating the distance of the perihelion.

162. Smart, D. "The Orbit of Halley's Comet." **Journal of the British Astronomical Association**, 16 (January 1906), 105-06.

Computes the orbit of the comet. Says that the alterations which he has computed may well change the date of any meteoric display from about 4 May to 17 May. See also entry 163.

163. Smart, David. "Halley's Comet." **Journal of the British Astronomical Association**, 16 (February 1906), 160-61.

Letter. Gives corrections for his January 1906 (see entry 162) paper, predicting that the comet will be well placed, but in the twilight sky--nevertheless it should be the brightest one we have seen in this (northern) hemisphere for twenty years.

164. Lynn, W.T. "Halley's Comet and 'Zadkiel'." **Journal of the British Astronomical Association**, 16 (April 1906), 238-39.

About Richard James Morrison of the Royal Navy who made astronomical predictions under the pseudonym, Zadkiel. In 1835 he published a pamphlet about "this potent heavenly messenger," predicting calamities before its actual apparition by invoking past events which coincidentally happened concurrent with the appearance of the comet. See entry 108.

165. Proctor, Mary. "Return of Halley's Comet in 1910." **Scientific American**, 94 (9 June 1906), 474.

Synoptic review of the comet's path, speed, and scheduled appearance. Notes that Halley's Comet is the first comet ascertained to move in an elliptical orbit.

166. Grover, C. "Halley's Comet, 1910 Return." **Journal of the British Astronomical Association**, 16 (July 1906), 353-54.

Presents a diagram of path of Halley's Comet through

stars at its 1910 return. Predicts a favorable position for the comet and therefore the probability of an early discovery.

167. "Halley and His Comet." **Scientific American,** 95 (4 August 1906), 87.

A romanticized telling of Halley's devotion to astronomical science and his prediction of the periodicity of the comets of 1531, 1607, and 1682.

168. Ledger, E. "Halley's and Other Comets." **Nineteenth Century,** 60 (September 1906), 379-95.

Discusses comets in general while devoting a quarter of the article to the discussion of Halley's Comet. Halley's is viewed as unique in its long traced history and many reappearances. In 1759 the comet's apparition tested for the first time the application of Newton's laws to astronomical bodies.

169. Lynn, W.T. "The Approaching Return of Halley's Comet." **Observatory,** 29 (November 1906), 432.

Because of a special combination of perturbations, the period between the last and the approaching return of the comet is the shortest since 1531, amounting to only 74 1/2 years. Lynn predicts that the comet will be visible to the naked eye in March 1910.

170. Crommelin, A.C.D. "Reply." **Journal of the British Astronomical Association,** 17 (December 1906), 96-97.

Replies to a query as to whether the return of the comet may be prevented by the same causes that prevented the Leonid display of November 1899. There is no parallel between the two cases; the comet will reappear.

1907

171. G[erard], J. "Of a Bull and a Comet." **The Month,** (February 1907), 151-57.

Recounts the story of Pope Callixtus III issuing a bull against the comet, then proceeds to show the apocryphal nature of the legend. The author provides quotes from some historians who have propagated the story.

172. Crommelin, A.C.D. "Abstract of Lecture Delivered Before the Association at the Meeting on February 27, On Halley's Comet." **Journal of the British Astronomical Association,** 17 (March 1907), 211-20.

Presents a review of historical events associated with the appearance of the comet. Crommelin suggests that the meteoric section of the Association keep a look-out for Aquarid meteors early in May each year from now until 1911.

173. Henkel, F.W. "Halley's Comet." **Knowledge and Scientific News,** (March 1907), 57-59.

Were the sun and comet alone in space the latter's orbit would be an exact ellipse. However this is not the case and the comet is affected by the forces of other heavenly bodies. Provides a discussion of the astronomical history of the comet as well as a diagram of its orbit.

174. "British Astronomical Association." **London Times,** 1 March 1907, 3.

Column f. A lecture by A.C.D. Crommelin at the meeting of the British Astronomical Association is reported. Crommelin tells of his study of the comet's perturbations. The comet's next return is expected on 16 May 1910. The watch for Halley's Comet should begin this year and continue till 1911.

175. Crommelin, A.C.D. "The Return of Halley's Comet in 1759." **Journal of the British Astronomical Association,** 17 (April 1907), 282-83.

Poems. Quotes from two poems extracted from **Gentleman's Magazine,** October 1758 and May 1759. Appended is a note stating that the first known American sighting of the 1759 apparition was at New York on the morning of 3 April 1759. See entry 18.

176. Henkel, F.W. "Halley's Comet." **Popular Astronomy,** 15 (April 1907), 238-45.

Gives some historical background on the comet. Most of the article is spent discussing factors involved in predicting its return.

177. Lynn, W.T. "Three Remarkable Comets." **Journal of the British Astronomical Association,** 17 (April 1907), 267-79.

Overview article on Encke's Comet, Newton's Comet, and Halley's Comet. Gives only brief attention to Halley's Comet, and that is devoted primarily to a synopsis of historical events concurring with its appearances.

178. Cowell, P.H. and A.C.D. Crommelin. "The Perturbations of Halley's Comet." **Monthly Notices of the Royal Astronomical Society,** 67 (12 April 1907), 366-411.

Presents lengthy tabular data and equations for

calculations of the perturbations by Jupiter between the comet's 1835 and 1910 returns. Preliminary calculation shows that comet will return about May 1910.

179. Henkel, F.W. "Halley's Comet." **Scientific American Supplement**, 63 (13 April 1907), 26154-55.

Gives a history of astronomical attention devoted to the comet. There is a diagram showing the orbit of the comet in relation the orbits of the planets in our solar system.

180. "Royal Observatory Greenwich." **London Times**, 10 June 1907, 4.

Column f. Mentions that preparations for observation of the comet are underway. It is appropriate that the Royal Observatory should observe the comet since this is where Halley's observations were made (actually Halley used Flamsteed's observations which were made at the Royal Observatory). Since Pontecoulant in 1864 no one else has published about the 1910 return of Halley's Comet.[sic]

181. Cowell, P.H. and A.C.D. Crommelin. "The Perturbations of Halley's Comet." **Monthly Notices of the Royal Astronomical Society**, 67 (14 June 1907), 511-21.

Presents lengthy tabular data and equations for calculations of the perturbations by Saturn between the comet's 1835 and 1910 returns.

182. Twain, Mark. "Extract from Captain Stormfield's Visit to Heaven." **Harper's**, 116 (December 1907), 41-49.

Short story. Part 2 of the story was published in **Harper's**, January 1908, pp. 266-76. Published as book in 1909, but actually written in 1868. The comet motif looms large in the story, with Halley's Comet mentioned twice (pp. 41 and 271) as a comparative reference.

183. Cowell, P.H. and A.C.D. Crommelin. "The Perturbations of Halley's Comet in the Past--First Paper: The Period 1301-1531." **Monthly Notices of the Royal Astronomical Society**, 68 (13 December 1907), 111-25.

Gives a meticulous description of the process by which the authors computed the perturbations of Halley's Comet for its 1301, 1378, 1456, and 1531 apparitions.

1908

184. Turner, H.H. **Halley's Comet: An Evening Discourse**

to the British Association, at Their Meeting at Dublin, on Friday, September 4, 1908. Oxford: The Clarendon Press, 1908. 32 pp.

A general discussion of Halley and his comet. Suggests that Halley pursued his calculations out of his enthusiasm for Newton's discoveries, rather than for the fame given him by posterity. Mentions historical events associated with the comet in the public mind.

185. Cowell, P.H. and A.C.D. Crommelin. "The Perturbations of Halley's Comet in the Past. Second Paper. The Apparition of 1222." Monthly Notices of the Royal Astronomical Society, 68 (10 January 1908), 173-79.

15 August 1222 is calculated as the date of perihelion passage (see also entry 183), showing Hind's indentification of the comet of July 1223 as Halley's to be in error (see entry 135). Explains complete computation and use of Williams' Chinese records (see entry 138).

186. Wendell, O.C. "Halley's Comet." Popular Astronomy, 16 (February 1908), 71.

An ephemeris is computed from Pontecoulant's of 1908--from October 1907 to May 1910 at three month intervals.

187. Cowell, P.H. and A.C.D. Crommelin. "The Perturbations of Halley's Comet in the Past. Third Paper. The Period from 1066 to 1301." Monthly Notices of the Royal Astronomical Society, 68 (13 March 1908), 375-78.

Corrects 1222 perihelion date to 10 September 1222 (see entry 185). The date of perihelion for 1066 is calculated to be 27 March considering vagueness of descriptions.

188. Cowell, P.H. and A.C.D. Crommelin. "The Perturbations of Halley's Comet, 1759-1910." Monthly Notices of the Royal Astronomical Society, 68 (13 March 1908), 379-95.

An investigation of the perturbations caused by the earth, Venus, and Neptune. The authors predict 8 April 1910 as the date of perihelion. They also predict that the comet will appear as a morning star till 17 March and after 11 May as an evening star.

189. Bailey, S.I. "Return of Halley's Comet." Science, 27 (27 March 1908), 512-13.

Halley's Comet having the effect of extreme interest, not only to the astronomer, but to the world at large, will soon appear. Presents some of the comet's astronomical history. Reprinted in **Scientific American Supplement**, (11 April 1908), p. 239.

190. Lynn, W.T. "Halley's Comet." **Observatory**, 31 (April 1908), 179.

Relates Flamsteed's observation of the comet from 20 August [31, NS] to 9 September [20, NS] 1682. Halley did see it, but did not actually observe it having recently been married and recently returned from Europe.

191. Cowell, P.H. and A.C.D. Crommelin. "Table Giving Approximate Values of the Perturbations of Halley's Comet by Jupiter and Saturn in the First and Fourth Quadrants of the Orbit." **Monthly Notices of the Royal Astronomical Society**, 68 (10 April 1908), 458-59.

Presents tabulations for the perturbations of Jupiter and Saturn as functions of the mean anomaly of the disturbing planet at the adjacent perihelion passage.

192. Denning, W.F. "The Meteors of Halley's Comet." **Nature**, 77 (30 April 1908), 619.

We possess no records of rich meteor showers having been observed in 1759 or 1835. The Aquarid meteor shower of May will at its radiant point conform very nearly both in date and place with particles following the path of Halley's Comet.

193. Wilson, H.C. "The Next Apparition of Halley's Comet." **Popular Astronomy**, 16 (May 1908), 265-70.

The period of the comet varies from 74 to 79.5 years. A table is presented giving approximate elements of Halley's Comet reduced to the equation of 1910. Also given is historical background on comet's apparitions and some explanation of the difficulty in exactly predicting its perihelion date.

194. Cowell, P.H. and A.C.D. Crommelin. "The Perturbations of Halley's Comet in the Past. Fourth Paper. The Period 760 to 1066." **Monthly Papers of the Royal Astronomical Society**, 68 (8 May 1908), 510-14.

Calculates 11 June 760, 28 February 837, and 19 July 912 as approximate dates of perihelion passage.

195. Dean, John Candee. "The Story of Halley's Comet." **Popular Astronomy**, 16 (June 1908), 331-45.

Reviews the superstition associated with comets, particularly Halley's Comet. Provides a diagram with course of Halley's Comet from 1 July 1908 to 1 October 1909. Says chances of the earth colliding with the comet are 1 in 280,000,000.

196. Lynn, W.T. "Halley's Comet." **Observatory,** 31 (June 1908), 254.

Letter. That Halley "discovered" the comet that bears his name is a common but erroneous idea. He "predicted" its periodicity.

197. Searle, George M. "The Impending Return of Halley's Comet." **Catholic World,** 87 (June 1908), 289-97.

Discusses the periodicity of the comet in layman's terms. Briefly reviews some of the historical events concurrent with the comet's appearance and the import of Halley's calculations.

198. Chambers, G.F. "Halley's Comet in 1458 and the Pope." **Journal of the British Astronomical Association,** 18 (July 1908), 379-81.

Follows the chronology of the story of Pope Callixtus III's supposed exorcism of the comet. It is traced to Platina's **Vitae Pontificum** (1479), which does not mention any bull, exorcism, or excommunication against the comet.

199. Holetschek, J. "Computed Magnitudes for Halley's Comet." **Nature,** 78 (2 July 1908), 207.

Abstracted from an article that appeared in **Astronomische Nachrichten,** 13 June 1908, p. 99. Monthly ephemerides showing probable positions of comet for October 1908 through March 1909 are provided. The author predicts perihelion passage for 16 May 1910.

200. Wilson, H.C. "The Next Apparition of Halley's Comet." **Scientific American Supplement,** 66 (5 September 1908), 158-59.

Diagram: "Halley's Comet and the Planets 1910." Also gives astronomical history of the comet with some comment about preparations for observation of the coming apparition.

201. "Evening Lectures." **London Times,** 7 September 1908, 4.

Column b. A lecture by H.H. Turner at Oxford on the subject of Halley's Comet is reported. The astronomical history of the comet and the story of Halley's calculations are synopsized. See also entry 184.

202. Rigge, William F. "The Pope and the Comet." **Popular Astronomy,** 16 (October 1908), 481-83.

Addresses the story of Pope Callixtus III and the papal bull supposedly issued against the comet. Rigge asserts that his research indicates that no such papal bull was ever issued and that the story is apocryphal, emanating from the Vatican librarian, Platina.

203. "Halley's Comet." **Popular Astronomy**, 16 (October 1908), 522.

Presented is a synopsis of the lore associated with Halley's Comet as well as a description of preparations underway at Yerkes Observatory for its observation. Quotes 1:281,000,000 as the odds of the earth colliding with comet.

204. Comstock, George C. et al. "Committee on Comets." **Astrophysical Journal**, 28 (November 1908), 410.

The announcement of the Astronomical and Astrophysical Society's appointment of a committee on comets in light of the approach of Halley's Comet. The committee's purpose is to ascertain what kinds of research should be undertaken at the present time--the subjects of research and the method to be employed in carrying it out.

1909

205. Cowell, P.H. and A.C.D. Crommelin. "The Perturbations of Halley's Comet in the Past. Fifth Paper. The Period B.C. 240 to A.D. 760." **Monthly Notices of the Royal Astronomical Society**, 68 (Supplement 1908), 665-70.

Perturbations by Venus, Earth, Jupiter, Saturn, and Uranus, and dates of perihelion for each of the apparitions between 240 B.C. and 760 A.D. are calculated.

206. Aitken, R.G. "Position Observations of Comet Halley." **Lick Observatory Bulletin**, 5 (1909), 165.

Observations with mean place for comparison stars (mid-October through mid-November 1909) are presented. On 17 November the comet was estimated to be a little brighter than a 12 1/2 magnitude star; the central condensation was very sharp, almost stellar.

207. Chambers, George F. "Halley's Comet." In **The Story of Comets, Simply Told for General Readers**. Oxford: Clarendon Press, 1909. pp. 102-25 and passim.

Appendix 4 is an ephemeris of Halley's Comet for January through July 1910. The sections on Halley's Comet were extracted and published as a separate,

Halley's Comet: With Brief Notes on Comets in General (Oxford: Clarendon Press, 1910), 47 pp. in order to take advantage of the market created by the interest generated in 1910 by the apparition of Halley's Comet (see entry 890). Chapter 9 is devoted entirely to Halley's Comet, presenting the comet's astronomical history as well as the many events associated with its apparitions. There is also discussion about the coming 1910 apparition.

208. Curtis, Heber D. "Position Observations of Comet Halley." **Lick Observatory Bulletin,** 5 (1909), 165.

Observations made 12 September through 15 November 1909 are presented. Positions are derived from plates made with the Crossley reflector at Lick Observatory.

209. Richards, L. Adolph. **Comets.** Winchester, VA: Eddy Press, 1909. 16 pp.

A pamphlet directed at the popular, marginally educated, market. Deals mostly with comet lore, superstition, and historical events associated with Halley's Comet.

210. Wright, W.H. "Note on the Spectrum of Halley's Comet." **Lick Observatory Bulletin,** 5 (1909), 146.

The spectrum of Halley's Comet was photographed on 22 October 1909 using the slitless spectograph attached to the Crossley reflector. The effective duration of exposure was two hours. There is no evidence of the existence of bright lines or bands.

211. Denning, W.F. "The Expected Rediscovery of Halley's Comet." **Observatory,** 32 (January 1909), 62–63.

Letter, dated 23 December 1908. Predicts that the comet will be discovered in September or October 1909, based on its discovery in 1758 77 days before perihelion passage and in 1835 101 days before perihelion. Halley's Comet varies its period from 74.88 to 79.34 years with mean being 76.8.

212. Smart, David. "Halley's Comet in 1910." **Journal of the British Astronomical Association,** 19 (January 1909), 121–24.

Brings information about Halley's Comet as up to date as possible. Presents a diagram of the orbit of the comet in 1910 on an equatorial plane. Predicts that meteors will radiate within a few days of 5 May.

213. Watson, A.D. "Halley's Comet and Its Approaching Return." **Journal of the Royal Astronomical Society of Canada,** 3 (January 1909), 210–19.

A poetic description of the comet and its formation. Presents an abbreviated chronology of historical events associated with the comet's past apparitions. Gives a schedule for the path and observation of the comet during the first half of 1910.

214. "Comets." **London Times,** 1 January 1909, 4.

Column c. Reporting of Dr. Smart's reading of a paper on "Halley's Comet in 1910." Since computing an ephemeris in 1901 from Pontecoulant's elements which made perihelion passage near the end of May 1910, astronomers have found errors in the calculation of the comet's eccentricity (see entry 212).

215. Lee, Oliver J. "The Photographic Search for Halley's Comet with the Two Foot Reflector of Yerkes Observatory." **Popular Astronomy,** 17 (March 1909), 160-61.

The comet was in the most favorable position for observation at about the time of the winter solstice. It shows traces of stellar light at the 17th or 18th magnitude.

216. Sampson, R.A. "Halley's Comet and Pope Callixtus." **Journal of the British Astronomical Association,** 19 (March 1909), 206.

His research shows that the source of the apocryphal story about Pope Callixtus III exorcising the comet is Laplace's **Exposition du Systeme du Monde** (1796).

217. Pickering, Edward C. "Comparison Stars for Halley's Comet." **Harvard College Observatory Circular,** No. 156, (30 March 1909), 3 pp.

Published as part of a bound volume by Harvard University Press, 1917. Presents a table of stars with which Halley's Comet may be compared and their magnitudes.

218. "Comet Notes." **Journal of the British Astronomical Association,** 19 (April 1909), 259.

All attempts to detect Halley's Comet during the winter proved unavailing, and the search must be abandoned until autumn. An ephemeris for the first half of 1910 is provided.

219. "Halley's Comet." **Edinburgh Review,** 209 (April 1909), 290-307.

Reviews articles by Cowell and Crommelin (see entries 183, 185, 187, 188, 191, 194, and 205) and books by Turner (see entry 180) and Williams (see entry 138) The world is waiting for news that the comet has been rediscovered upon a photographic plate. After looking back at the history of the comet, the attitude of expectation is summarized: "It is clear, however, that the physicist is not quite ready to step in at the point where the dynamical astronomer confesses that he must stop, content perforce with his result that something, connected with the sun, happens sometimes to a comet and releases it from its obedience to Newton's gravitation laws. ... We await, then, the return of Halley's Comet with an interest which is intensified by the sporting chance that the prediction of Messrs. Cowell and Crommelin may after all be in error."

220. ["Note."] **Observatory**, 32 (April 1909), 186.

An anonymous correspondent draws comparison between the Norman conquest of England in 1066 and a predicted German invasion of England in 1910, pointing to the coincidences of the comet, both aggressors being named William, and both dates containing a 10.

221. Crommelin, Andrew C.D. "The Expected Return of Halley's Comet." **Science Progress**, 3 (April 1909), 543-57.

A review of the comet's history, providing a list of 29 perihelion dates, the number of days between successive returns, where seen, and notes. Nearly half of the comet's period (from Dec. 1856 to April 1889) was spent on small arc of its orbit lying beyond the planet Neptune.

222. Lynn, W.T. "Halley's Comet in 1835." **Observatory**, 32 (April 1909), 175-77.

Discusses various 1835 observations, and points out that the Europeans have failed to take into account the observations of Olmsted and Loomis at Yale (see entry 121), the only place in America with an observatory from which accurate readings of heavenly bodies could be taken in 1835.

223. Grover, C. "Path of Halley's Comet, 1909-1910." **Journal of the British Astronomical Association**, 19 (July 1909), 394.

Three diagrams are presented depicting the path of the comet through the stars from 12 September 1909 to 23 June 1910.

224. Matkiewitsch, L. ["Note"]. **Journal of the British Astronomical Association**, 19 (July 1909), 408.

An ephemeris with the date of perihelion in elements given as 16 April 1910 at 6:00 A.M. Reprinted from **Astronomische Nachrichten**, no. 4295.

225. Larkin, Ralph B. "The Approach of Halley's Comet." **North American Review**, 190 (August 1908), 194-99.

During recent years astronomers have been looking forward with high expectancy to the comet's 1910 apparition. Presented is some comet lore and pseudo-scientific discussion about the nature of comets. Larkin misinforms that Halley plotted the course of the comet as he observed it, when in fact he used Flamsteed's observation data for his calculations (see also entry 190).

226. Metcalf, Joel H. "Searching for Halley's Comet." **Popular Astronomy**, 17 (August 1908), 440-42.

Four attempts were made to photograph the comet at the observatory at Taunton, Massachusetts. All attempts failed. The procedures used are described.

227. Roberts, Alexander W. "Halley's Comet." **Scientific American**, 101 (14 August 1908), 110.

A diagram of Halley's Comet during its cycle 1835-1910. Described are the path and schedule of the comet. "Thus if there are planets beyond Neptune they will make their presence felt in disturbing the comet as it passes its aphelion goal." The comet is eagerly awaited.

228. "Expecting Halley's Comet." **New York Times**, 25 August 1909, 6.

Column 2. Observatories worldwide scan for the first sighting of Halley's Comet. There is a keen rivalry among astronomers for this honor.

229. Heward, Ed. Vincent. "The Story of Halley's Comet." **Nineteenth Century**, 66 (September 1909), 509-26.

The comet is closely associated with events that have molded the destiny of Europe. Heward asserts that Halley first caught sight of comet on his way to Paris, then observed it with Cassini. Reprinted in **Living Age**, 263 (9 October 1909), 67-81.

230. Serviss, Garrett P. "The Coming Spectacle in the Skies." **Hampton's Magazine**, 23 (September 1909), 381-88.

"A thrill of apprehension will be felt by millions who see its fiery head plowing through the stars ... as if monitory of unknown and unavoidable evil." Gives a brief account of the comet's astronomical history. Plays upon the public's fear of a chance of collision between the comet and the earth.

231. "Halley's Comet Seen." **New York Times**, 13 September

1909, 1.

Column 4. Professor Wolf at Heidelberg Observatory is the first to sight Halley's Comet.

232. "Comet Has Famous History." **New York Times,** 14 September 1909, 1.

Column 2. A summary of the comet's history, "It encouraged William the Conquerer and evoked a Papal Bull."

233. "The Return of Halley's Comet." **London Times,** 11.

Column c. "It is seldom that a telegram of such interest to astronomers is circulated" as that just received from Professor Wolf in Heidelberg announcing the rediscovery of Halley's Comet. Also provides synopsis of the comet's history.

234. Touchstone. "The Coming of the Comet." **London Daily Mail,** 15 September 1909, [p. ?].

Poem.

235. "Sees Comet Through Lens." **New York Times,** 18 September 1909, 8.

Professor S.W. Burnham of Yerkes Observatory, Williams Bay, Wisconsin, reports sighting the comet through a telescope without the aid of camera.

236. "Position of Halley's Comet." **New York Times,** 22 September 1909, 1.

Professor E.B. Frost of Yerkes Observatory gives a detailed position of the comet.

237. Seaman, Owen. "To Halley's Comet (Shortly Expected In Our Neighborhood)." **Punch,** 22 September 1909, 200.

Poem. Reprinted in **Living Age,** 263 (30 October 1909), 305.

238. "The Return of Halley's Comet." **Scientific American,** 101 (25 September 1909), 208.

Comments on Wolf's discovery of the comet at Heidelberg and how much this discovery was coveted by astronomers everywhere. The comet made its appearance 4 months before expected. Gives a brief history of the comet.

239. "Halley's Comet Brighter." **New York Times,** 28 September 1909, 1.

Column 6. Professor E.E. Barnard of Yerkes

Observatory observed the comet on the 24 September, finding it considerably brighter than it was on the 17th. It is estimated as being of the 15th magnitude, with indefinite condensation almost amounting to a small nucleus without definite boundaries.

240. "Comet Notes." **Observatory**, 32 (October 1909), 399-400.

The long expected Halley's Comet has been detected. Relates various sightings and presents an ephemeris for 11 September 1909, 8:40 P.M., Greenwich Mean Time.

241. "Comet Notes." **Journal of the British Astronomical Association**, 20 (October 1909), 51.

The comet is increasing in brightness--mentions several observations. It is small, about 15 seconds in diameter, well defined, with central condensation. Perihelion is predicted for 19 April 1910 at 7:00 A.M. GMT. Presents a corrected ephemeris for April and May 1910.

242. "Halley's Comet." **Journal of the British Astronomical Association**, 19 (October 1909), 456.

Corrects ephemeris to indicate perihelion on 20 April 1910. Calculates the apparent magnitude of the comet to be 11.7 at the end of 1909. Gives elements.

243. ["Comet Notes"]. **Observatory**, 32 (October 1909), 402-03.

Presents following elements of orbit predicated on perihelion passage of 20 April 1910; longitude of ascending node; node to perihelion; inclination of orbit; semi-major axis of ellipse; and eccentricity. When comet is in perihelion its heliocentric longitude will be about 306 degrees.

244. Lee, Oliver J. "Photographs of Halley's Comet." **Astrophysical Journal**, 30 (October 1909), 237-38.

Observed at Yerkes Observatory 16, 17, 24, 26 September 1909. The first photograph was taken just four days after Wolf's initial sighting at Heidelberg; cloudy weather had precluded any earlier photographs. Reprinted in the **Journal of Royal Astronomical Society of Canada**, 3 (Sept. 1909), pp. 341-43.

245. Raulein, Theodore M. "Splendid Visitor from Afar." **Technical World**, 12 (October 1909), 146-51.

A brief review of historical lore associated with Halley's Comet. Presents a table of returns of the comet from its perihelion in 240 B.C. through 1910

A.D. There is a synoptic explanation of the comet's periodicity. It is inaccurately stated that it was first sighted in 1758 by a big telescope.

246. "Halley's Comet." **New York Times**, 1 October 1909, 7.

Column 1. The Yerkes Observatory photographs the comet which is now 350 million miles away. In addition to being the first photograph ever taken in the United States of the comet, this is believed to be the first reproducible photograph of the comet ever taken anywhere.

247. "Will See Halley's Comet." **New York Times**, 2 October 1909, 1.

Column 2. Seagrave predicts that the comet will be visible to the naked eye on the 19th of May 1910 when the earth will be "swept by star dust." Seagrave calculates that at its nearest approach the comet will be 6,235,000 miles away from earth.

248. "Halley's Comet." **Yukon Sun** [Yukon, Oklahoma], 8 October 1909, 1.

Column 1. It is a monster in size and it is probable that its tails will sweep the earth. This may cause a meteoric shower, such as occurred in olden times, but no damage is expected ... the soothsayers will get busy and predict the end of time.

249. "The Pope and the Comet." **Scientific American,** 101 (9 October 1909), 259.

Letter. A reader corrects misinformation previously published (see entry 238) in the magazine regarding the apocryphal story of Pope Callixtus III and the comet.

250. Campbell, W.W. "Return of Halley's Comet." **Publications of the Astronomical Society of the Pacific,** 21 (10 October 1909), 188-94.

Presents an abbreviated history of the comet. There is an explanation of the comet's nucleus, coma, and tail. Notes that astronomers welcome the opportunity to use the photo-dry plate and spectroscope to remove some mystery of comets. 4 plates. Reprinted as **Smithsonian Institution Report,** 1909, pp. 253-59; also reprinted as a separate as **Smithsonian Institution Publication** no. 1958.

251. Curtis, Heber D. "Observations of Halley's Comet." **Publications of the Astronomical Society of the Pacific,** 21 (10 October 1909), 211.

The search for the comet had begun in early September

1909 with the Crossley reflector. Provides observation data for 12, 13, 14, and 22 September.

252. Kaempffert, Waldemar. "Halley and His Comet." **Outlook**, 93 (23 October 1909), 428-34.

An historical rendition of Halley's prediction of the comet's periodicity. A lay description of the comet and its orbit (with diagram of orbit) is provided. "A blind sportsman, bent on duck-shooting, stands a better chance of hitting his target than the earth of ramming a comet."

253. "Halley's Comet Seen Again." **New York Times**, 25 October 1909, 6.

Column 6. E.E. Barnard of Yerkes Observatory reports sighting the comet on both the 17th and 19th of October. The magnitude of the comet is not less than 13.5, it is about 15 seconds in diameter with an indefinite brightening in the middle and no elongation.

254. Newall, H.F. "To the Editor." **London Times**, 25 October 1909, 6.

Column c. Letter. The rediscovery of Halley's Comet in the position predicted by Cowell and Crommelin "demonstrated in an extraordinary way the refinement of astronomical prediction in the hands of the masters." The comet's magnitude is estimated to be 14 to 14.5.

255. "Halley's Comet Brighter." **New York Times**, 27 October 1909, 1.

Column 6. Harvard Observatory professors report seeing the comet with the aid of telescopes. The comet appears to be growing brighter somewhat rapidly. It was seen without difficulty on 17 October with the 24" reflector at the Harvard Observatory.

256. "Comet Notes." **Journal of the British Astronomical Association**, 20 (November 1909), 104-05.

Halley's Comet brightened up during November at an unexpectedly rapid rate--it is now estimated to be at the 10th magnitude. An ephemeris for September through November 1909 and an ephemeris for December 1909 through March 1910 are presented based on a perihelion date of 19 April 1910.

257. "Photographs of Halley's Comet." **Popular Science**, 75 (November 1909), 518-19.

A brief review of observations in 1909 to date and the telescopic equipment used. Included are two recently taken photographs of the comet.

71

258. "Return of the Most Famous Comet In the Universe." **Current Literature,** 47 (November 1909), 558-61.

A diagram of the relative motions of Halley's Comet and the earth from May to July 1910. There is also a synoptic review of several magazine articles dealing with historical events associated with the comet's apparitions. Mention is also made of some of the astronomical attention being devoted to the comet during its present return.

259. Crommelin, A.C.D. "Notes." **Observatory,** 32 (November 1909), 435-36.

Relates further observations of the comet and presents a corrected ephemeris computed on the basis that perihelion will be 19 April 1910. Communicates that the American Comet Committee is sending Ellerman to Honolulu in order to increase the chances for observation.

260. ["Note"]. **Observatory,** 32 (November 1909), 444-45.

List of historical event identifications that concur with appearances of Halley's Comet. The list is reprinted from the **London Daily Mail,** 16 September 1909.

261. ["Report of Meeting"]. **Journal of the British Astronomical Association,** 20 (November 1909), 70-72.

Reports of observations and plans for observation. Interesting insight into the Association's attention to the comet in preparation for its perihelion passage.

262. Phillips, T.E.R. "Observations of Halley's Comet With a 12 1/4-inch Reflector." **Journal of the British Astronomical Association,** 20 (November 1909), 94-95.

Observations made on 16, 21, 22, 23 November 1909. Viewing not very good.

263. Roberts, Alexander W. "Halley's Comet." **Chambers's Journal,** 81 (November 1909), 710-13.

The feeling one has waiting for the return of Halley's Comet is held to be similar to waiting for a ship or homing swallows to return. Gives a brief astronomical history of the comet and poses some of the questions that scientists hope to answer this time around. For the scientist the comet is viewed as an object of extreme importance.

264. Willis, Edgar C. "Halley's Comet, 1909." **Journal of the British Astronomical Association,** 20 (November 1909), 95-97.

Observed at Norwich, England on 25 October; 5, 13, 15, 20 November; and 5 and 8 December 1909.

265. "Halley Comet to Cross Sun." **New York Times**, 5 November 1909, 1.

Column 4. Calculations by Father Searle show that the comet will cross the sun's face on 18 May at 9:15 P.M.

266. "Earth in Path of Comet." **New York Times**, 6 November 1909, 1.

Column 5. Professor E.C. Pickering says that the earth will pass through Halley's Comet on 18 May. The tail is gaseous--acetylene, carbonic, and several others. The gas is rarified so will not be perceived by the average earthling. If gases were apparent to humans the result would be very "disagreeable."

267. Cowell, P.H. and A.C.D. Crommelin. "Note on the Time of Perihelion Passage of Halley's Comet." **Monthly Notices of the Royal Astronomical Society**, 70 (12 November 1909), 3-4.

The authors calculate the time of perihelion passage to be April 19.65, and the time of transit of the comet over the sun as 18 May, visible in the Pacific Ocean, Asia, and Australia, etc. They offer the suggestion that the spectroheliograph and short exposure photograph will be the most effective means of observation.

268. Jacoby, Harold. "The Coming of Halley's Comet." **Harper's Weekly**, 53 (20 November 1909), 13.

Halley's Comet has come back to our solar system again. The earth will pass through the comet's tail, but astronomers are able to confidently assert that there is no danger to the earth or its inhabitants. The earth has passed through a comet's tail once before and nobody noticed.

269. "Return of Halley's Comet: An Ephemeris for This Year." **Scientific American Supplement**, 68 (27 November 1909), 351.

Presents an ephemeris for November and December 1909. Perihelion is predicted for 20 April 1910 and the transit of the sun's disk for 18 May 1910. Reprinted from **Knowledge and Scientific News**.

270. "Comet's Tail Around Us." **New York Times**, 28 November 1909, 1.

Column 6. Professor J.A. Brashear says that the comet's tail will encircle the earth about 18 May but will not harm us. Speaking before the Outlook Club in New York City on 27 November he showed for first time in United States the photograph of the comet taken by Wolf at Heidelberg.

271. "Comet Notes." **Journal of the British Astronomical Association,** 20 (December 1909), 158-59.

Reports observations from Barcelona, Melbourne, Meudon, and Mt. Hamilton. Also reported is that the nucleus appeared double on 8 December 1909.

272. "Notes." **Observatory,** 32 (December 1909), 476-77.

Halley's Comet has brightened considerably during November at an unexpectedly rapid rate. It will be within the reach of small instruments in December and January. It will pass its ascending node on 17 January and its descending node on 18 May 1910.

273. "The Course of Halley's Comet." **Popular Mechanics,** 12 (December 1909), 716-17.

Provides a diagram giving the relative course of the comet and the earth from 15 May 1909 to 22 July 1910. Shows the comet to be nearest the earth on 10 May 1910.

274. Andrews, E.C. "The Danger of the Comet." **Pearson's Magazine,** 28 (December 1909), 610-18.

Panic mongering of the worst kind in view of the approach of Halley's Comet. This article recounts past calamities concurring with comet's appearance, vividly describing (with pictorial accompaniment) the disasterous effect of either a collision or close brush with the comet.

275. Barnard, E.E. "Suggestions In Respect to Photographing Comets , With Special Reference to Halley's Comet." **Popular Astronomy,** 17 (December 1909), 597-609.

An excellent "how to" guide for the astronomer and non-astronomer alike. Provides an insight into state-of-the-art astronomical photographic equipment and technique.

276. McKready, Kelvin. "The Comet in the Christmas Sky." **Ladies Home Journal,** 27 (December 1909), 39.

Contrasts the "pleasure and curiosity" with which we are greeting the current appearance of Halley's Comet with the "alarm, popular terror, social hysteria, and stupendous horror" aroused among men of old by such visitations. Some astronomical history provided as well.

277. Phillips, Theodore E.R. "Halley's Comet." **Journal of the British Astronomical Association,** 20 (December 1909), 155-56.

Letter. On 22 November the comet was about magnitude 10, and has not since regained its brightness. The comet has usually exhibited the appearance of a small stellar nucleus surrounded by nebulous haze.

278. Searle, George M. "Present Probabilities About the Comet." **Catholic World,** 90 (December 1909), 289-93.

Discusses the scheduled appearance of comet based on the computations of Cowell and Crommelin. The visibility of the comet cannot be predicted because of several variables. Its tail will point directly toward us, as comet tails always point away from the sun.

279. Warner, Irene E. Toye. "Great Events in the World During Apparitions by Halley's Comet." **Knowledge and Scientific News,** (December 1909), 463-66.

Starting with B.C. 467 through A.D. 1835 the author recounts calamities, battles, and deaths that occurred relatively concurrent with the apparitions of Halley's Comet.

280. Scomp, Henry Anselm. "Halley's Comet and Solomon's Temple." **Independent,** 67 (2 December 1909), 1253-56.

Conjectures that the comet appeared in 1088 B.C., suggesting that Halley's Comet was the "angel with a flaming sword, between heaven and earth" seen by King David over Mt. Moriah (II Samuel 2, I Chronicles 21). David therefore chose Moriah as the home of the Ark of the Covenant.

281. Gore, J.G. "Edmund Halley: The Man Who Dispelled Cometary Superstition." **Scientific American Supplement,** 68 (4 December 1909), 363.

A short biographical sketch of Halley, focusing on his prediction of the periodicity of what has become known as Halley's Comet.

282. Hallock, William. "Halley's Comet: Celestial Wanderer Again Nears the Earth." **New York Times,** 5 December 1909, V, 3.

Its visits are traced for twenty centuries, primarily from an astronomical viewpoint. Illustrations are included.

283. Dunelm, Handley. "Halley's Comet." **Yorkshire Post,** 7 December 1909, [p.?].

Poem.

284. Aitken, R.G. "December Observations of Halley's Comet." **Publications of the Astronomical Society of the Pacific,** 21 (10 December 1909), 259.

Observations of the comet with a 36-inch refractor telescope on the nights of 14, 15, 17, 18 December 1909. The comet appeared to be nearly round, with a sharp nucleus. There was no evidence of any tail. Brightness is estimated to be of the 10th magnitude.

285. Curtis, H.D. **Publications of the Astronomical Society of the Pacific,** 21 (10 December 1909), 259-61.

The comet is rapidly increasing in brightness. It will be closest to the earth on 20 May 1910 at 14,300,000 miles. The probability is that the earth will pass through the comet's tail on 18 May.

286. Phillips, T.E.R. "Note on the Variation in Brightness in Halley's Comet." **Monthly Notices of the Royal Astronomical Society,** 70 (10 December 1909), 183.

Observation with 12.5-inch Calver reflector telescope shows some considerable variation in the comet's brightness. The comet was most easily seen under inferior conditions on 22 November 1909, but was especially difficult to see on 6 December when conditions were very good.

287. "Halley's Comet." **Scientific American,** 101 (11 December 1909), 436.

The comet was observed at Smith Observatory, Geneva, New York on 27 November 1909 during a total eclipse of the moon. It was moving westerly through Taurus at the rate of one degree daily, but changing very slightly in declination toward the south.

288. "The New Comet." **London Times,** 14 December 1909, 6.

Column c. Reports the discovery of a new comet (not Halley's). Also provides positions and observation instructions for Halley's Comet.

289. Barnard, E.E. "Visual Observations of Halley's Comet." **Astronomical Journal,** 26 (22 December 1909), 43-44.

Observations made with the micrometer of a 40-inch telescope, 17 September through 30 November 1909, at Yerkes Observatory. Two tables are presented showing the position of the comet and comparison stars. A daily description for each observation of the comet is provided.

290. "Halley's Comet Misshapen." **New York Times,** 25 December 1909, 1.

   Column 6. Professor Percival Lowell's photograph taken at Flagstaff, Arizona shows the comet to be of irregular formation, but with a distinct nucleus, which is out of the center.

291. "Suggested Observations of Halley's Comet." **Nature,** 82 (30 December 1909), 260-62.

   Gives suggested techniques for observation based on the circular sent to observatories by the Comet Committee of the Astronomical and Astrophysical Society of America--photographic, spectroscopic, photometric, and polariscopic methods are addressed.

1910

292. Aitken, John. "Did the Tail of Halley's Comet Affect the Earth's Atmosphere?" **Proceedings of the Royal Society of Edinburgh,** 30 (1910), 529-50.

   Read 18 July 1910. The author carried out dust count experiments in the Northwest Highlands of Scotland. Nothing was found to prove that the cause of the increase in dust and haze during June and July 1910 could be attributed to Halley's Comet. No colorful sunsets were observed.

293. Aitken, R.G. "Observations of Halley's Comet." **Lick Observatory Bulletin,** 6 (1910), 74-75.

   Observations made December 1909 through July 1910, giving mean places of comparison stars.

294. Barton, Samuel G. **Halley's Comet.** Potsdam, NY: by the author, 1910. 23 pp.

   This lecture presents a most balanced compendium of general astronomical background for Halley's Comet. Headings within the essay suggest its focus: importance of superstition; appearance; number, mass, volume, etc.; light; the tail; danger from comets; origin; discovery; orbits; the elements of an orbit; Halley's work on comet orbits; all comet orbits approximately parabolas; perturbations; application to Halley's Comet; variation in appearance; history; present return; discovery; apparent path in the sky; results expected; Edmund Halley. Provides various diagrams to illustrate points made in text. Reprinted from the **Clarkson Bulletin,** v. 7 (January 1910), 5-24. The author is professor of mathematics at the Thomas S. Clarkson Memorial School of Technology, Potsdam, N.Y.

295. Borgmeyer, Charles J. **Halley's Comet: A Lecture Delivered at St.   Louis, Mo., January 25, 1910.** St. Louis, MO: St.   Louis University, 1910.   43 pp.

Presents astronomical history of the comet, nicely spiced with some comet lore. Of particular interest is a detailed refutation of the apocryphal story of Pope Callixtus III issuing a papal bull against the comet. Includes sections devoted to envelopes and tail, radiation pressure, paths of comets, conic sections, the elements, concluding with remarks as to what is going to happen during the present apparition. "Our probable or possible encounter with the tail next May is burdened with so many IFS that it seems hardly worth while to say more about it." Published as volume 6, No. 1 of the **Bulletin of St.   Louis University.**

296. Brown, John. **Halley's Comet: Its History, With That of Other Noted Comets, and Other Astronomical Phenomena, Superstitions, etc.** London: Elliot Stock, 1910.   52 pp.

Contains nothing new, but collects existing facts and references about the comet in a manner palatable to the general public.   The four illustrations are rather crude and of no special interest.

297. Cowell, P.H. and A.C.D. Crommelin. **Essay on the Return of Halley's Comet.** Leipzig: W.   Engelmann, 1910. 60 pp.

A long essay presenting the authors' calculations predicting the exact date of perihelion and location of the 1910 apparition of Halley's Comet. The essay was awarded the A.F. Lindemann Prize of the Astronomische Gesellschaft (see entry 156 for announcement of prize).

298. Cowell, P.H. and A.C.D. Crommelin. **Investigation of the Motion of Halley's Comet from 1759-1910.** Edinburgh: Neill, 1910.   84 pp.

The authors build on their previous work presented in the **Monthly Notices of the Royal Astronomical Society** (see entry 188).

299. Crawford, R.T. and W.F. Meyer. "Orbit of Halley's Comet." **Lick Observatory Bulletin,** No.   6 (1910), 7.

Presents the comet's orbit and calculation upon which it is based.

300. Crommelin, A.C.D. "Address on the Return of Halley's Comet in 1910." **Journal of the Transactions of the Victoria Institute,** 25 (1910), 18-34.

Presented at the annual general meeting of the Philosophical Society of Great Britain, on 9 May 1910, this is an entertaining and interesting astronomical history of the comet through its various recorded apparitions. There is argument that the nucleus is comprised of solid matter, in that it appears to move exactly as if by the force of gravity alone. See entry 505 for a **London Times** report of the lecture, and see entry 509 for a report by the **New York Times**.

300A. Dix, Irving Sidney. "The Comet." in **The Comet and Other Verses**. Carbondale, Pennsylvania: Press of Munn's Review, 1910. pp. 7-8.

Poem.

301. Edwards, Gus and Harry B. Smith. "Mr. Earth and His Comet-Love." 1910.

Song. Inspired by people's fixation with Halley's Comet. Sung in the 1910 Ziegfeld Follies. Music by Edwards and lyrics by Smith are reprinted in **Ziegfeld Follies 1910**, Harry B. Smith, ed. 1911.

302. Elson, Henry William. **Comets: Their Origin, Nature and History**. New York: Sturgis and Walton, 1910. 54 pp.

A discussion of comets in general with special reference to Halley's Comet, directed at a popular audience. Chapters are : the solar system and the stars; comets and their orbits; superstitions about comets; some remarkable comets; meteors and shooting stars; and the last chapter (pp. 46-54) is devoted specifically to the popular history of Halley's Comet. Includes a plate of the comet in 1835 and a diagram of the comet's 1835 to 1910 cycles, plotting locations in relation to the earth and sun.

303. Emerson, Edwin. **Comet Lore: Halley's Comet in History and Astronomy**. New York: Schilling Press, 1910. 127 pp.

A sensational rendition of the terror that the comet has inspired throughout history. Terms Halley's Comet as the "bloodiest of all" comets based on historical concurrence of events. The final two chapters address the "peril from collision" and the "end of the world."

304. Zwack, George M. **The Return of Halley's Comet and Popular Apprehensions**. Manila: Bureau of Printing, 1910. 22 pp.

Dispells in as empirical terms as possible fears about the possibility of the comet colliding with the earth or the sun. "We may look for all kinds of wild rumors circulated by ignorant or malicious persons, setting forth in glowing colors all the horrors which the 'comet' is going to bring upon us poor mortals. It is

believed that the present discussion of these alleged
dangers, if given the widest publicity possible, will
help in allaying the nightmares of the timid and
furnish material to those who desire to dispel the
fears of their neighbor."

305. Findlay, James. "Halley's Comet." **Chambers's
Journal**, 82 (February 1910), 255-56.

Letter. Points out historians' errors with various
calendars. Using Aristotle's description of a comet in
371 B.C. (370) calculates 370, 294, 218, 142, 66 B.C.,
10, 86, 162, 238, 314, 390, 466, 542, 618, 694, 770,
846, 922, 998, 1074, 1150, 1226, 1302, 1378, 1454,
1530, 1606, 1682, 1758, 1834, 1910, and 1986 A.D.
"Pretty close isn't it?"

306. Hicks, E. Rupert. **Philips' Model of Halley's Comet,
Earth, and Sun.** London: G. Philip and Son, 1910.

Designed by Hicks and sold at one shilling. The comet
and the earth are given on a scale of one inch to about
20 million miles. The model is like a planisphere and
fails to give any notion of an accurate idea of the
respective paths of the earth and comet. Its distorted
scale leaves a false impression of the distance of the
nucleus from the earth. At the end of one pointer is
the earth and at the end of another the comet, the
latter with an appendage which represents the tail.
The two paths are marked with months and days, so that
by moving the two pointers to the place of any given
day we can see the relative position of the two bodies.

307. **The Story of Halley's Comet.** Boston: George P. Ide
and Co., 1910. 4 pp.

A four page booklet reprint of an article originally
appearing in the February 1910 issue of **Circle
Magazine.** Presents a brief history of the various
appearances of Halley's Comet from 240 B.C. to 1910
A.D.

308. Slipher, V.M. and C.O. Lampland. "Preliminary
Notes on Photographs and Spectrographic Observations of
Halley's Comet." **Lowell Observatory Bulletin,** 1, No. 47
(1910), 252-54.

Covers the period of 14 April to 1 May 1910 inclusive,
with the exception of 28 April. On no two consecutive
mornings did the comet present the same appearance on
the negatives, though on different dates there are
similarities.

309. Thomson, Arthur. "On a Variation in the Intensity of
the Penetrating Radiation at the Earth's Surface Observed
During the Passage of Halley's Comet." **Proceedings of Royal**

Society of Canada, (1910), 61-65.

Observations on the intensity of radiation when the comet and tail were close to earth is described. The experiment used an ionization vessel of about 30,000 ccs capacity which was installed on the roof of a building; readings were taken with a quadrant electrometer.

310. Todd, David. **Halley's Comet.** New York and Cincinatti: American Book Co., 1910. 16 pp.

This pamphlet was put out as part of a series of educational pamphlets from the publisher; it also served as an advertisement for Todd's book, **Todd's New Astronomy,** pages 392 through 411 of which (the chapter dealing with comets and meteors) have been excerpted and appended to the cited pamphlet. Directed at a popular audience, there is general discussion of the comet's astronomical history, some comet lore, and a rendition of historical events associated in the public's mind with the comet. Presents a list of places in the sky where the comet may be observed on various dates, 1 January through 24 May 1910.

311. ["Report of Meeting"]. **Journal of the British Astronomical Association,** 20 (January 1910), 184-85.

Reports communications received by A.C.D. Crommelin regarding the observation of the comet.

312. "Comet Notes." **Journal of the British Astronomical Association,** 20 (January 1910), 215-16.

Observations reported from Bath, Ealing, and Abingdon in England, and a note that D. Ross is the first to sight comet in Australia on 18 November 1909.

313. ["Notes"]. **Observatory,** 33 (January 1910), 67-68.

Refers to circular issued by the Astronomical Society of America's Committee on Comets giving advice on best way to photograph Halley's Comet. Takes umbrage that the British community of astronomers were not consulted. The tone of this article exhibits an intense sense of rivalry, perhaps even petty jealousy.

314. Broderick, W.B. "Halley's Comet from the Norman Point of View." **Science Progress,** 4 (January 1910), 492-503.

Quotes from contemporary annals and concludes that the comet, inspiring the Normans and terrifying the English, "was a potent factor in determining the outcome of the Battle of Hastings."

315. Campbell, Rev. Frederic. "The Religious
Significance of Halley's Comet." **Homiletic Review,** 59
(January 1910), 10-13.

The author is president of the Dept. of Astronomy at
Brooklyn   Institute,   N.Y.   Campbell discusses the
probabilities of the earth colliding with a comet and
concludes that if it is God's will it will not hit us.

316. Doolittle, C.L. "The History of Halley's Comet."
Popular Science, 76 (January 1910), 5-22.

An overview of the history and the observation of the
comet through various apparitions. Well illustrated
with photographs, drawings, and charts.

317. Eddie, L.A. "Halley's Comet." **Journal of the British
Astronomical Association,** 20 (January 1910), 202-03.

Observations taken at Grahamstown, England for 10-15
January 1910.   There is no sign whatever of a stellar
nucleus. The comet exhibits a slightly condensed
center and fades off imperceptibly outwards, so as to
render the periphery indefinable.

318. MacDonnell, W.J. "Halley's Comet in a Small
Telescope." **Journal of the British Astronomical
Association,** 20 (January 1910), 200.

Read at meeting of the New South Wales (Australia)
Branch of the Association, 17 December 1909. Halley's
Comet appeared ill-defined and elongated, at times a
stellar-like point was suspected. Condensation at its
southern end was unmistakable.

319. Merfield, C.J.   "Halley's Comet." **Journal of the
British Astronomical Association,** 20 (January 1910),
201-02.

An abstract of a letter to W.J. MacDonnell.   Presents
a corrected ephemeris for September through November
1909. States a suspicion that the comet actually
varies in brightness.   Predicts that the earth will
probably pass through the tail of the comet.

320. Mitchell, S. Alfred. "Evening Sky Map for January."
Monthly Evening Sky Map, 4 (January 1901), 1, 3.

The first page presents a diagram of the approximate
path of Halley's Comet from 11 September 1909 through
25 May 1910. Of the comets with a period of less than
80 years Halley's is the only one that is known to move
with a retrograde motion.

321. O'Neill, H.C. "The Coming Comet." **Windsor Magazine,**
31 (January 1910), 231-35.

82

A pastiche of astronomical lore and history. Included is a diagram of the present course of Halley's Comet in relation to the greater planets; there is also a reproduction of the comet as it appears in the Bayeux tapestry. At perihelion the comet, it is estimated, will be travelling 3 to 4 million miles in a day.

322. Reeve, Arthur B.   "Halley's Homing Comet." **Scrap Book**, 9 (January 1910), 103-05.

Gives historical events associated with the comet from 66 A.D. through 1758. Closes with summary of Halley's discovery of the comet's periodicity.

323. Tebbutt, J. "Note on Halley's Comet." **Journal of the British Astronomical Association**, 20 (January 1910), 199-200.

Originally read at meeting of the New South Wales (Australia) Branch of the Association, 17 December 1909. Predicts perihelion to be on 20 April 1910 and an arrival of its descending node on 18 May at 10:26 P.M. GMT.

324. Dean, John C. "Relative Positions of Halley's Comet, the Earth, and the Sun." **Scientific American,** 102 (8 January 1910), 27.

Good brief description of the path and scheduled apparition of the comet in relation to observability from earth; a diagram is included.

325. Dunn, Frederic Stanley. "The Julian Star." **Classical Weekly,** (8 January 1910), 87.

Erases the misperception that the comet appearing at the time of Julius Caesar's death was Halley's. Not so, Halley's nearest appearance was probably 11 B.C., thirty years after the assassination. See also entry 367.

326. "Tail to Halley's Comet." **New York Times,** 11 January 1910, 1.

Column 6. Yerkes Observatory reports seeing a tail trailing behind the comet's nucleus. The comet may now be seen through a small telescope as a faint nebula. It is too faint to be caught by anything except a highly sensitized photographic plate.

327. Crawford, Lord. "Halley's Comet." **London Times,** 13 January 1910, 11.

Column d. Letter. Warns of political unrest in the world. "As portents of nature make the opportunity for the agitator, it behoves [sic] all civilized countries by foresight to guard against such possible dangers ..." (i.e., political upheaval).

328. "Havana Sees Halley's Comet." **New York Times,** 15 January 1910, 5.

Column 3. (blurb-one column inch). The Bolen Observatory reports that on 13 January the comet was sighted in the form of a faint white cloud between the planet Mars and the O star of Pisces.

329. Mayer, Hy. **New York Times,** 16 January 1910, V, 16.

Cartoon. A befuddled sitting man is questioned by a pipe smoking friend who looms over him: "Its all up, Chester, its all up./What's happened?/Astronomers say Halley's Comet has no tail!"

330. "Discussions at the British Astronomical Association." **London Times,** 27 January 1910, 10.

The new comet and Halley's Comet were discussed at last night's meeting. Crommelin said that Halley's Comet might develop a tail at any time. It can be seen by the public toward the end of April on through May. Americans sent an observer, Ferdinand Ellerman, to Sandwich Islands in the hope of bettering the chances for observation.

331. Roberts, Ruel W. "How an Amateur May Find Halley's Comet." **Scientific American,** 102 (29 January 1910), 102 ff.

The author's ambition was to be the first amateur astronomer to discover Halley's Comet. He found it using 3.5-inch telescope. Roberts describes how the amateur can ascertain where to look for the comet.

332. "Comet Notes." **Journal of the British Astronomical Association,** 20 (February 1910), 273.

A continuation of the ephemeris for Halley's Comet for June and July 1910 computed by F.E. Seagrave.

333. "Ephemeris of Halley's Comet." **Publications of the Astronomical Society of the Pacific,** 22 (February 1910), 37-38.

An ephemeris is presented, assuming perihelion passage to be 19.65 April 1910.

334. Cortie, A.L. "The Devil, the Turk, and the Comet." **Observatory,** 33 (February 1910), 91-95.

About the long held parallel connection between the fall of Constantinople and the 1456 passage of the comet--points out that in reality the city fell in

1453. Goes on to dispell the myth about the comet's excommunication by Pope Callixtus III.

335. Crommelin, A.C.D. ["Notes"]. **Observatory,** 33 (February 1910), 104-05.

Questions the deduction of new elements by C.J. Merfield (see entry 319). Calls attention to a curious historical parallel between the 1835 and 1910 British general elections, drawing comparisons between Liberal and opposition membership in Parliament.

336. Curtis, H.D. "Halley's Comet." **Publications of the Astronomical Society of the Pacific,** 22 (February 1910), 33.

The comet continues to increase in brightness, though not so rapidly as during November and the first part of December. Photographs show the tail as a narrow cone and later as several fine streamers from the head. A plate (negative taken with Crossley reflector, 5 February 1910) is included.

337. Curtis, Heber D. "Note on Photographs of Halley's Comet." **Lick Observatory Bulletin,** 5 (February 1910), 183.

The negatives of photographs taken with the Crossley reflector on 11-13 December 1909 show the brighter portion of the coma and faint traces of a short, cone-shaped tail. Negatives taken on 4 February 1910 shows a very fine, sharp, stellar nucleus.

338. Eddie, L.A. "Halley's Comet." **Journal of the British Astronomical Association,** 20 (February 1910), 247-48.

Observations taken at Grahamstown, England for 26, 29, 31 January and 2, 6, 13 February 1910. The comet is observed becoming perceptibly brighter on the 29th and 31st of January 1910. On February 6th it has suddenly lost its tail.

339. Mitchell, S. Alfred. "Evening Sky Map for February." **Monthly Evening Sky Map,** 4 (February 1910), 3.

The spectroscope is to be the important instrument of investigation for Halley's Comet, since the comet promises to be the first bright comet to visit the earth since the instrument has been developed--it will allow the tail to be deciphered.

340. Olivier, Charles P. "Physical Ephemeris of Halley's Comet." **Publications of the Astronomical Society of the Pacific,** 22 (February 1910), 34-35.

An ephemeris for January through April 1910 is calculated on the basis of sunrise and sunset and the sun being 12 degrees below the horizon at morning and evening at Mt. Hamilton. Perihelion date is assumed to be 19 April.

341. Whitmell, C.T. "Halley's Comet." **Journal of the British Astronomical Association,** 20 (February 1910), 248-51.

Computations of the comet's orbit are explained and presented. The nearest approach of the comet to earth is calculated to occur on the 19th and 20th of May at 13.75 million miles. See note in March 1910 issue, pp. 317-18.

342. Cowell, P.H. "On Halley's Comet As Seen from Earth." **Nature,** 82 (3 February 1910), 400-01.

A tail of 20-30 degrees in length is expected. A diagram gives the position of the earth for 6 days in May and the position of the comet on 27 dates measured from perihelion passage; there is also a table with elliptical coordinates. Reprinted **Scientific American Supplement,** 19 March 1910, p.188.

343. Rutledge, Archibald. "To Halley's Comet." **Youth's Companion,** 83 (3 February 1910), 60.

Poem.

344. "Comet's Poison Tail." **New York Times,** 8 February 1910), 1.

Column 4. Yerkes Observatory discovers that spectra from Halley's Comet show the tail to contain poisonous cyanogen gas. Cyanogen is a very deadly poison, a gram of its salt touched to the tongue would be enough to cause instant death. Based on this discovery there is disagreement about the degree of danger posed to humans.

345. "No Danger from Comet." **New York Times,** 10 February 1910, 1.

Column 6. W.J. Hussey, director of the observatory at the University of Michigan says there is no danger from cyanogen gas in comet--"not enough poison to kill an insect." He ridicules the notion that earth-life is endangered by the comet.

346. "Poison in the Tail of a Comet." **New York Times,** 11 February 1910, 10.

Column 4. Editorial. Even though cyanogen is a terrible poison, the tenuity of the comet's misty tail poses no danger. Only to the ignorant and the superstitious is the comet alarming.

4. Cartoon by Hy Mayer, 16 January, 1910. See entry 329. © 1910 by The New York Times Company. Reprinted by permission.

347. "Ephemeris of Halley's Comet." **Scientific American,** 102 (12 February 1910), 143.

Computed at Goodsell Observatory, Carleton College, Northfield, Minnesota. During December 1909 the comet became bright enough to see with telescopes.

348. "Notes on Halley's Comet." **Scientific American Supplement,** 69 (12 February 1910), 107.

Gives interesting facts about "our celestial visitor" including its observed diameter, heliocentric conjunction in longitude, and a diagram for the night of 18 May 1910 when the earth will be immersed in the comet.

349. **New York Times,** 13 February 1910, III, 4.

Announces that the paths of the earth and Halley's Comet will cross on 18 May. A diagram shows paths and intersection.

350. "Balloon View of Comet." **New York Times,** 15 February 1910, 1.

C.J. Glidden of Boston plans balloon ascent from Boston to photograph the comet on about 1 May from an altitude of three miles.

351. Hughes, C.E. "How to See Halley's Comet." **Punch,** (16 February 1910), 117.

Provides satirical advice on how not to miss seeing the comet.

352. Taylor, C.S. **Nature,** 82 (17 February 1910), 499-500.

Points out that the altitude of the sun at North Cape on May 18th, 1910 will be 1 degree 9', and the contact does not take place until 4:06 P.M. local time (GMT).

353. **London Times,** 18 February 1910, 6.

Column e. Oxford University is to confer degrees on P.H. Cowell and A.C.D. Crommelin for services rendered in the prediction of the exact date of Halley's Comet's return to perihelion. See entry 297 for the essay containing their calculations and prediction.

354. "Halley's Comet." **Scientific American,** 102 (19 February 1910), 162.

A diagram of the apparent path of Halley's Comet from 5 January to 5 April 1910. Gives measurements as reported in other periodicals. Reprinted from **Nature.**

355. "Halley." **Saturday Evening Post,** 19 February 1910, 47.

Advertisement. Silver Brand Collars with lincord buttonholes introduces two collar heights, Halley at 2 1/8 inches, and Comet at 3 3/8 inches. Ruth Freitag (see entry 1279) states that this ad also appeared widely in college newspapers.

356. "Professor Turner on Halley's Comet." **London Times,** 19 February 1910, 11.

Reports on H.H. Turner's address at the Royal Institution last night. Turner saw Halley's Comet as appealing on historical and sentimental grounds rather than as a really magnificent object of and in itself. Also provides some of the historical associations which make it of such interest. See entry 962 for text of the address.

357. Chalmers, Stephen. "The Truth About the Comet." **New York Times,** 20 February 1910, V, 12.

Satirical story about the disappearance of a Dr. Gook, Heer Flounder, and the truthful author. "It should be clear then that the comet known as A-1910 is not a new comet, but merely the Halley Comet which Heer Flounder towed out of its natural orbit. ..."

358. Barnard, E.E. "Observations of Halley's Comet and Notes on Its Photographic Appearance." **Astronomical Journal,** 26 (21 February 1910), 76-77.

Gives a description of the various aspects of the comet at intervals of observation, January through 10 February 1910.

359. "Halley's Comet." **London Times,** 24 February 1910, 10.

Gives a lay account of the comet and its approach. "There can be little doubt that Halley's Comet has a considerable amount of solid matter in it, for a mere bunch of vapour could not hold together for even a short time, much less 2000 years."

360. Lowell, Percival. "The Coming Comet." **Youth's Companion,** 83 (24 February 1910), 99-100.

Presents an astronomical history and explanation of the comet geared at a level for the adolescent youth. Pictures and diagrams are included. The opportunity to see the comet is rare. Practically speaking, like Halley, one only sees it once.

361. "Halley's Cometary Studies." **Scientific American Supplement,** 69 (26 February 1910), 143-44.

Reviews Halley's studies of comets, placing the
periodic comet that became know as Halley's Comet in
this larger context.

362. "A Comet in a Suitcase." **New York Times,** 27 February
1910, 10.

Column 4. Editorial. Cambridge Observatory director,
Sir R.S. Ball, comments on the comet by quoting
Herschel that "the whole comet could be squeezed into a
portmanteau." The **New York Times** thinks it better to
leave it where it is, it is safer there than here on
earth.

363. "Comet Notes." **Journal of the British Astronomical
Association,** 20 (March 1910), 325-26.

The comet was observed as an evening star until the
middle of March. It would have been readily visible to
the naked eye if it could have been seen up on a dark
background. A recalculated ephemeris is given for
20-31 May 1910.

364. "Notes." **Terrestrial Magnetism,** 15 (March 1910),
37-38.

Reports synopses of communications from various German
astronomers relating their preparations for the
observation of Halley's Comet.

365. Chant, Clarence A. "Halley's Comet." **Journal of the
Royal Astronomical Society of Canada,** 4 (March 1910),
104-05.

A history for the general reader, first published in
the February 1910 issue of **Westminster** (Toronto).
Discusses Halley's calculations toward the discovery of
the comet's periodicity and gives some historical
events often associated in the public mind with the
comet's appearance.

366. [Crommelin, A.C.D.] ["Notes"]. **Observatory,** 33
(March 1910), 142-43.

Presents a continuation of the ephemeris of Halley's
Comet for June and July 1910 as computed by F.E.
Seagrave. The comet will hardly be able to be followed
in Europe after July. See also p. 150 for the lyrics
(doggeral) for a song done to the air of "Sally in the
Alley."

367. Dunn, Frederic Stanley. "Julian Star." **Popular
Astronomy,** 18 (March 1910), 164-65.

Disproves the misperception that it was Halley's Comet
which appeared at the time of the death of Julius

Caesar. Dunn calculates that the closest date of apparition of Halley's Comet was in 11 B.C., 30 years later. See also entry 325.

368. Hirayama, K. "Halley's Comet in Japanese History." **Observatory,** 33 (March 1910), 130-33.

A translation of the records appearing in the Nihonsyoki or Nihongi, the earliest chronicle of Japan, for the years coinciding with Halley's Comet 684 A.D. through 1222 A.D.

369. Phillips, T.E.R. "Observations of Halley's Comet." **Journal of the British Astronomical Association,** 20 (March 1910), 309-10.

Observations between 16 November 1909 and 5 March 1910 taken with a 12.25-inch Calver equatorial telescope on thirty nights. The comet was seemingly much less affected by the zodiacal light than might be anticipated. The increase in its brightness was more uniform and rapid than expected.

370. Pickering, William H. "Halley's Comet: Suggestions for Its Observation." **Popular Astronomy,** 18 (March 1910), 129-32.

One cannot expect that the light of the sun will be appreciably dimmed upon the transit of the comet across it. The head of the comet consists of a swarm of meteors and a small amount of extremely rarefied gas.

371. Rigge, William F. "The Apparent Path of Halley's Comet in the Sky." **Popular Astronomy,** 18 (March 1910), 165-69.

The comet moves in an elongated ellipse. There is a diagram of the orbit relative to the four inner planets; also included is a diagram of the apparent path of the comet through the stars. The comet is visible only in the morning before sunrise.

372. "Could the Earth Collide with a Comet?" **Scientific American,** 102 (5 March 1910), 194.

Editorial. Completely dispels possibility of the earth's colliding with Halley's Comet. Chances are 1 in 40,000,000 of being struck by the core of a visible comet and 1 in 4,000,000 of being struck by some part of the nucleus.

373. "Where Will It Strike? A Comet That Has Cut Loose in the Republican Constellation." **Puck,** 67 (9 March 1910), [8-9].

Cartoon. The comet, labeled "ALLDS Investigation," is

colliding with an exploding planet, many other planets pictured, each with the face of a Republican Party notable.

374. "Halley's a Twilight Comet." **New York Times**, 10 March 1910, 1.

Column 2. The most favorable time to look for the comet is between 6:30 and 7:30 P.M. with glasses elevated somewhere above where the sun sets. There is no danger from the comet unless it comes in contact with large quantities of hydrogen--and that is all locked up in water.

375. "Halley's Comet." **Spectator**, 103 (12 March 1910), 414.

The fact that the comet is not yet conspicuous does not mean that it will not be so in a few weeks' time. Developments in astronomical photography will make observation of this apparition of Halley's Comet infinitely more valuable. Gives some brief history of Halley's calculation of the comet's periodicity and the lore associated with its apparitions.

376. "Comet's Tail a Vacuity." **New York Times**, 13 March 1910, III, 3.

On account of its vacuity, the tail could have no injurious effect on humanity. The director of Lowell Observatory calls the comet's tail the "airiest approach to nothing set in the midst of naught."

377. "Collisions with Comets." **Literary Digest**, 40 (19 March 1910), 536-37.

Reaffirms that nothing harmful will happen as a result of the earth being "immersed" in the tail of Halley's Comet. Quotes from an editorial in the **Scientific American**, 5 March 1910 (see entry 372). Included is an illustration of the Bayeux tapestry and a cartoon of Kaiser Wilhelm and the comet.

378. Cowell, P.H. "Halley's Comet as Seen from the Earth: The Position of the Comet." **Scientific American Supplement**, 69 (19 March 1910), 188.

On May 18th the earth will probably be in the tail of the comet. A diagram gives the position of the comet and earth in relation to each other for 10, 14, 18, 22, 26, and 30 May 1910. Cowell suggests that apparitions which we regard as unrecorded were really recorded and the records subsequently lost.

379. "Comet A Is Vanishing." **New York Times**, 20 March 1910, 1.

But Halley's Comet grows in grandeur and will appear in April. When it emerges from behind the Sun in late April it is expected to be a spectacle in the morning skies.

380. Lynn, W.T. "Halley's Comet." **London Times,** 22 March 1910, 13.

Column b. Letter. Points out that Halley's Comet is in a certain position very soon after noon, and is too near the sun to be visible, therefore a reporting correspondent must be in error in his supposed sighting of the comet.

381. Griggs, H.W. "Condensed Facts About Halley's Comet." **Scientific American,** 102 (26 March 1910), 258.

Two diagrams depicting the apparent path of Halley's Comet through the heavens and Halley's Comet and the earth. Given are the speed and predicted schedule for the comet along with some historical associations.

382. Pickering, Edward C. "Brightness of Halley's Comet." **Harvard College Observatory Circular,** No. 157 (30 March 1910), 3 pp.

Published in bound volume by Harvard University Press, 1917. Results of measures of the comet's nucleus made with a photometer, with achromatic prisms attached to a 15-inch equatorial telescope. The comet apparently underwent change due to the heat of the sun.

383. "Comet Notes." **Journal of the British Astronomical Association,** 20 (April 1910), 387-90.

A synopsis of various observations. Presents a brightness table and an ephemeris for August through 6 December 1910.

384. "The Return of Halley's Comet." **Bulletin of the American Geographical Society,** 42 (April 1910), 261-65.

If the earth came in contact with comet's tail, less than .0000001 % of it would enter the earth's atmosphere. Chances of comet colliding with earth are less than that of the earth's falling into the sun, since earth's diameter is less than .01 of the sun's.

385. "Transit of Halley's Comet." **Proceedings of the Astronomical Society of the Pacific,** 22 (April 1910), 101-02.

Refers to C.S. Taylor's article in **Nature** (see entry 352). Also cites recent calculations that comet's times of ingress and egress for the Pacific Slope are: 6:22 A.M. PST and 7:22 A.M. PST respectively on 18

May 1910. Transit will be invisible in Europe and much of America.

386. ["Notes"] **Observatory,** 33 (April 1910), 190.

Poem. Reprinted from **Washington** [D.C.] **Star** titled "The Comet." Also clippings from the St. **James Gazette** (9 March 1910) and the **Chicago Tribune.**

387. ["Report of Meeting"]. **Journal of the British Astronomical Association,** 20 (April 1910), 345.

Crommelin reports on literature and observation of Halley's Comet during the last month.

388. Buzzell, Francis. "The Approach of Halley's Comet." **Popular Mechanics,** 13 (April 1910), 491-94.

Explains what must happen in relation to the comet, the earth, and the sun in order for the earth to pass through the tail of Halley's Comet. A diagram of the orbits of the earth and comet, and the position of sun when the tail may sweep the earth is presented.

389. Campbell, W.W. "The Coming of Halley's Comet." **Sunset,** 24 (April 1910), 365-76.

An overview article giving synopsis of the relationship between Newton and Halley, Halley's prediction of the comet's periodicity, describes the basic parts comet and behavior of a comet. Also a diagram, Herschel's 1835 drawing of the comet, and a reproduction of the Bayeux tapestry panel with the comet.

390. Crawford, R.T. "Note on the Orbits of Comets Halley; a1910, and e1909 (Daniel)." **Publications of the Astronomical Society of the Pacific,** 22 (April 1910), 97-98.

The elements for the comet are given--computed from observations of 17 September and 16 December 1909, and 28 February 1910. Halley's Comet furnishes an example of a very long arc in the case of a nearly parabolic orbit.

391. Crommelin, A.C.D. "Notes." **Observatory,** 33 (April 1904), 182-83.

Presents a recalculation of the ephemeris for the last ten days of May by Dr. Smart using Merfield's elements. Mentions that the comet was seen at Heidelberg on 11 February with an opera glass.

392. Curtis, H.D. "Note on Halley's Comet." **Publications of the Astronomical Society of the Pacific,** 22 (April 1910), 96-97.

The comet at the end of April was fairly conspicuous in spite of strong moonlight, rising two hours ahead of the sun. The effective magnitude of the comet is estimated to be 1.8 and the tail could be seen for 8-10 degrees. The edges of the tail are much brighter than is the center.

393. Dotson, H.P. "Another Who Saw Halley's Comet in 1835." **Popular Astronomy**, 18 (April 1910), 256.

Letter. The author was twelve years old in 1835. He then mistook the comet's transit across the sun for a buzzard flying. The comet is remembered as having covered 1/20th of the whole disk of the sun. Also mentioned are other astronomical phenomena which accompanied the apparition of Halley's Comet in 1835.

394. Eddie, L.A. "Halley's Comet." **Journal of the British Astronomical Association**, 20 (April 1910), 363-64.

Observations are reported for 14th through 16th April 1910. The nucleus of the comet was bright, equal to a 2.8 magnitude star. The star and faint traces of coma could be detected preceeding the comet. The tail was in greater evidence than had been expected.

395. Lee, William Ross. "To Halley's Comet." **Munsey's Magazine**, 43 (April 1910), 12.

Poem.

396. Merfield, C.J. "Method of Finding the Correction of the Predicted Time of Perihelion Passage of a Planet or Comet, and Its Application to Finding This Correction for Halley's Comet." **Journal of the British Astronomical Association**, 20 (April 1910), 365-67.

Presents the formula for computation and a table of coefficients from which curves were constructed for the comet September through November 1909.

397. Mitchell, S. Alfred. "Evening Sky Map for April." **Monthly Evening Sky Map**, 4 (April 1910), 3.

The cover presents a diagram of the paths of Halley's Comet and the earth about the sun. Positions of the comet are given for April and May. The month of April will be the most important in the history of Halley's Comet.

398. Pickering, William H. "The Return of Halley's Comet." **Century Magazine**, 79 (April 1910), 909-17.

A good general article, listing some curious historical coincidences and superstitions. Discusses the weight and constitution of comets, and categorically rules out the possibility of a collision between Halley's Comet and the earth.

399. Rigge, William F. "An Historical Examination of the Connection of Calixtus [sic] III with Halley's Comet." **Popular Astronomy**, 18 (April 1910), 214-19.

A complete review of the controversy over whether Pope Callixtus III actually excommunicated or exorcised the comet. He did not. Evidence against is taken primarily from the recently published work of Father Stein of the Vatican Observatory.

400. See, T.J.J. "The Return of Halley's Comet and Its Passage Between the Earth and Sun on May 18." **Munsey's Magazine**, 43 (April 1910), 3-12.

A discussion of the astronomical history of Halley's Comet and then an explanation of what will happen during the 1910 apparition. The comet will be the most striking phenomenon of this kind during the present generation.

401. Smith, Lucia E. "The Comet." **Sunset**, 24 (April 1910), 371.

Poem.

402. Kaempffert, Waldemar. "The Most Famous of Comets." **Collier's National Weekly**, 45 (2 April 1910), 21-22.

Gives a stylized history of the comet and events associated with it. The air we breath is as dense as iron in comparison with the diaphanous thinness of a comet's tail. We will be unaffected. Photographs of the comet and of the telescope at Yerkes Observatory are included.

403. Wilson, H.C. "Approach of Comet." **Northfield News** [Northfield, Minnesota], 2 April 1910, 1.

The author is director of the Goodsell Observatory at Carleton College. Presented is a history of the comet, a chart of its orbit, and a chart of its path across the sun's disk on 18 May 1910. Viewing with the naked eye or perhaps aided by field glasses is suggested.

404. "Watch of the Comet." **Baylor County Banner** [Seymour, Texas], 8 April 1910, 10.

Advertisement. "The red dragon of the sky. Watch children for spring coughs and colds. Careful mothers keep Foley's Tar and Honey in the house. Its prompt use has saved many lives." This ad ran in each issue through May.

405. Fouts, L.F. "Where to Look for Halley's Comet." **Baylor County Banner** [Seymour, Texas], 8 April 1910, 3.

From Trinity Mills, Texas. The comet will on 4 April first become visible to the naked eye. Gives dates and times of rising and setting for April and May.

406. Rambaut, A.A. "Places of Halley's Comet, 1909-1910, Deduced from Photographs Taken at Radcliffe Observatory, Oxford." **Monthly Notices of the Royal Astronomical Society,** 70 (8 April 1910), 497-502.

Presents a table of places of the comet for 7 November 1909 through 11 February 1910. Explains methodology for determining places via photographic and visual observation. Also included are the observer's remarks.

407. "Halley's Comet Reappears." **New York Times,** 9 April 1910, 1.

Column 4. Emerging from the glow of the sun's disk the comet is not yet visible to the naked eye. It will probably be two weeks before it can be seen without a telescope.

408. "The Comet's Intentions." **Literary Digest,** 40 (9 April 1910), 693.

Reiterates that the comet will have absolutely no harmful effect on the earth. Quotes at length in translation from an article by Edward Guillaume in **Le Nature,** 26 February 1910.

409. Chambers, George F. "Halley's Comet I: The Most Interesting of Periodical Comets." **Scientific American Supplement,** 69 (9 April 1910), 238-39.

Presents the astronomical history of the comet through the 1835 apparition. Continued in the 16 April issue (see entry 417).

410. "See Halley's Comet Again." **New York Times,** 11 April 1910, 7.

Column 2. Naval Observatory, Washington, D.C. sights the comet which is visible for the first time in 75 years. Seen at 4:00 A.M. with a 26-inch telescope.

411. "The Comet and the Sleepyheads." **New York Times,** 12 April 1910, 10.

Column 3. Editorial. It is a harmless, gaseous thing that does not even know that there is such an insignificant planet as earth in existence. Yet people in Indiana attribute sleepiness to it because its poisonous gases get into the blood.

412. **Illustrated London News**, 13 April 1910, d.

Advertisement. Uses a drawing that would appear in an article about the comet in the 23 April issue (see entry 435) to advertise itself: "There is No Doubt that the Comet will Not Strike the Earth; and there is No Doubt that the ... ILLUSTRATED LONDON NEWS Strikes the World Each Week."

413. "Wisconsin Observatory Sees Comet." **New York Times**, 13 April 1910, 20.

Column 4. (blurb-1 column inch) Yerkes Observatory sees comet. Only the body was visible as the tail is obscured by atmospheric disturbances.

414. "Halley's Comet and Composition of Its Tail Perplexing Problem for World's Astronomers." **Minneapolis Morning Tribune** [Minneapolis, Minnesota], 14 April 1910, 13.

Astronomers at the University of Minnesota are scouring the heavens for their first sight of the comet's blazing trail. The milkman will no doubt be the most popular man, awakening patrons to see the comet. A diagram of comet's course is included.

415. "Halley's Comet." **Nature**, 83 (14 April 1910), 201.

Mentions observations at Perth, Vienna, and Cape Town, as well as the terror being caused by the comet in southern Russia, and the suicide of a Hungarian farmer. It is announced that the Manila Weather Bureau has put out a brochure (see entry 304) in which supposed catastrophes associated with comet are shown to have no relation whatsoever to the comet.

416. "Students Gaze Nightly Through Telescope for Glimpse of Comet." **Minneapolis Morning Tribune** [Minneapolis, Minnesota], 15 April 1910, 9.

The observatory is one of the most popular places at the University of Minnesota these days. Students are anxious to get a glimpse of Halley's Comet and its brilliant tail. Included is a photograph of the observatory and an inset of a student looking through a telescope.

417. Chambers, George F. Halley's Comet--II: By Far the Most Interesting of the Periodic Comets." **Scientific American Supplement**, 69 (16 April 1910), 255-56.

Continued from 9 April issue (see entry 409). Carries forth the discussion to predictions for 1910 apparition and then turns discussion to historical observation and associations prior to 1682.

5. Cartoon by Hy Mayer, 17 April, 1910. See entry 420. © 1910 by The New York Times Company. Reprinted by permission.

418. Kaempffert, Waldemar. "The Peril of the Comet." **Saturday Evening Post**, 182 (16 April 1910), 3 ff.

A panic mongering article under the guise of science and history. Presents a discussion of the question, "if the earth were struck by a comet's core?" and concludes that it would be reduced to a cinder.

419. Russell, Henry Norris. "Halley's Comet At Its Brightest." **Scientific American**, 102 (16 April 1910), 317-18.

Presented is a diagram of the relative positions of the orbit of Halley's Comet, the earth, and the sun; also photographs. Whatever may be the origin of the intrinsic light of comets, it is responsible for most of the phenomena that make them of general interest--it is not reflected sunlight.

420. Mayer, Hy. **New York Times**, 17 April 1910, V, 16.

Cartoon. A sign painter is superimposed on the heavens where each planet has sign announcing Halley's Comet and large comet has its nose poking through a sign reading "tailless wonder." Caption: "The best advertised show in the world."

421. "Earth to Burn Path Through Tail of Comet." **Minneapolis Morning Tribune** [Minneapolis, Minnesota], 19 April 1910, 7.

"Sky traveller moves 31 miles every second." Presents a calendar (19 April - 5 June) for the comet. A map is also provided showing the comet's path of travel. There is a reaffirmation that people need not fear the hydro-carbon gas in the comet's tail.

422. "Comet Is Getting Near." **New York Times**, 20 April 1910, 1.

Column 6. Observatories (Willemstad, Curacao; St. John's, Newfoundland; Paris; and Rome) report that the comet is nearer. It is visible to the naked eye in Curacao. In southern Italy there are reports that people are showing great anxiety over the comet.

423. Leverin, Albert. "Faster! Faster!! Faster!!!" **Puck**, 67 (20 April 1910), [5].

Cartoon. Caption reads "New View of Halley's Comet, Showing the Souls of Lost Motorists." The comet is sailing through space with a steering wheel mounted atop the head with man in a driving coat and goggles in the driver's seat; other passengers are in the same attire.

424. "Halley's Comet." **London Times**, 21 April 1910, 13.

Column d. It is now possible to see the comet before sunrise, though it is difficult owing to its nearness to the sun. The comet is now nearly exactly stationary and therefore Venus is often mistaken for it. There is a comparison of position to 1145 apparition.

425. "Halley's Comet." **Nature**, 83 (21 April 1910), 223-25.

Two diagrams showing condition of observation, plus ephemeris for April and May 1910 are presented as are photographs of the comet. Reports from China state that the comet is being used as an omen to influence rioters. To allay public fear and alleviate the political situation, the government has posted circulars that exhibit facts about earlier appearances of Halley's Comet.

426. ["Notice"]. **Nature**, 83 (21 April 1910), 230.

Announcement of a cardboard model of the comet's path relative to the sun and earth for 10 March through 30 May 1910. The cost is one shilling one pence and it is recommended as a method of illustrating to those non-astronomical persons the comet's periodic appearance.

427. Birkeland, K. "The Transit of Halley's Comet Across Venus and the Earth in May," **Nature**, 83 (21 April 1910), 217-18.

Letter. Relates observation of the comet from northern Norway, 7 May through 1 June 1910. Suggests that it is conceivable that the tail of the comet may consist chiefly of electrical corpuscular rays, thereby being sensitive to the earth's magnetic impulses.

428. Eastman, Elaine Goodale. "Halley's Comet." **Independent**, 68 (21 April 1910), 865.

Poem.

429. "Comet Gazers Are Now Busy." **Carson City Daily Appeal** [Carson City, Nevada], 22 April 1910, 1.

Comet gazers are as thick as flies and on every street corner of the city there is a human with his head elevated skyward and neck stretched to a marked degree endeavoring to point out to another the exact location of the heavenly visitor.

430. "Halley's Comet Seen Just Before Sunrise." **Daily Iowan** [Iowa City, Iowa], 22 April 1910, 1.

Column 3. A brief description of the comet's

anticipated schedule. Local authorities scoff at the idea of the comet going to pieces with disasterous results. However, they admit that this was the fate of Comet Biela which disappeared by flying into bits.

431. "Harvard Sees the Comet." **New York Times**, 22 April 1910, 14.

Column 2. The Harvard Observatory sights the comet. Halley's Comet in its present stage hardly repays early rising. It appeared at 3:48 A.M. and disappeared from sight at 4:12 A.M. The magnitude of the nucleus is estimated to be about 6.4 with total brightness around the 5th magnitude.

432. "Chicagoans Fear Comet." **New York Times**, 23 April 1910, 1.

Column 4. Women in street cars have become hysterical and the population in the parts of the city inhabited by immigrants attribute the darkness over Chicago to the comet. Halley's Comet is creating a general consternation among the more ignorant.

433. "Dr. Bryant Sights Comet." **Minneapolis Morning Tribune** [Minneapolis, Minnesota], 23 April 1910, 1.

Dr. Bryant is the first Minneapolitan to sight Halley's Comet--or, at least, is the first to announce it. The comet was seen as a bright tailless object, 40 degrees above the Eastern horizon. The good doctor feels that he has one-upped the University astronomers.

434. "Mark Twain and Halley's Comet." **New York Times**, 23 April 1910, 10.

Column 6. Letter. Calls attention to the parallel between Twain's life span and the appearances of Halley's Comet. Twain was born 30 November 1835 (the comet's perihelion was 16 November 1835) and he died 21 April 1910 (the comet's perihelion was 21 April 1910).[sic] Fifteen days is the total difference between Twain's lifespan and the two perihelion dates.

435. "Reassuring for Fearful Chinese: Why Halley's Comet and the Earth Will Not Collide." **Illustrated London News**, (23 April 1910), 597.

A diagrammatic drawing showing the body of the comet with a two-million-mile long tail crossing the path of the earth. A little comet history is presented. Those who believe the earth is in danger are urged to study this diagram to see impossibility of a collision.

436. "The Comet Breeding Dissensions." **Literary Digest**, 40 (23 April 1910), 805-06.

"To future generations it will seem as if Halley's Comet this year had fallen among astronomic experts like the proverbial bone of contention among ravenous dogs." Quotes from various disagreeing articles.

437. "Big Demand for Telescopes." **New York Times**, 24 April 1910, II, 11.

Column 2. Because everybody wants to look for the comet sales of telescopes during the past three months have exceeded those for the entire period back to Civil War. Wholesalers have exhausted supplies, even trying to buy back instruments from local retailers.

438. "Making Preparations for a Comet's Visit." **New York Times**, 24 April 1910, V, 2.

A drawing shows how the sky over New York City will look when the comet is brightest. The comet will approach to within 14 million miles of the earth.

439. "Astronomers Suspected." **New York Times**, 25 April 1910, 5.

Column 3. Letters have flooded into the Harvard Observatory. Letter writers hint that astronomers have invented Halley's Comet or are hiding it. Many letters ask if there is a way to avoid the comet or if there is any cure for it. Still, others say that there is no comet because the earth is flat.

440. "To Study Comet's Effects." **New York Times**, 25 April 1910, 5.

Column 3. The United States government asks wireless operators and mariners to note any effects the comet may have on transmission or reception. The appearance of meteorites should also be logged by watch officers. See also entry 804.

441. "Halley's Comet." **London Times**, 26 April 1910, 10.

Column b. The comet is observed at the Royal Observatory at Greenwich. It is estimated to be of the second magnitude in brightness. Also mentioned are observations from Chile and Spain.

442. Innes, D.M. "To the Editor of the Times." **London Times**, 26 April 1910, 10.

Column b. Letter. Halley's Comet was plainly visible with the naked eye from 4:00 A.M. till about 4:30. The nucleus was extremely brilliant, but even with good field glasses the tail could not be seen. It was low down in eastern sky.

443. "Comet's Tail Gone." **New York Times**, 27 April 1910, 1.

Column 6. (blurb-one column inch) Zurich, Switzerland reports comet visible to naked eye for 55 minutes. No trace of tail is seen, even with telescope.

444. "Wireless Won't Feel Comet." **New York Times**, 27 April 1910, 1.

Column 6. Professor O.C. Wendell of the Harvard Observatory expects no wireless interference to occur due to the comet.

445. "Flashes from Comet's Tail." **New York Times**, 28 April 1910, 2.

The Rev. L.A. Harvey tells the New York City Unitarian Club that he saw the head of the comet from his home in Flatbush appearing as bright as Mars.

446. "Halley's Comet." **Nature**, 83 (28 April 1910), 259-60.

A synopsis of various observations from around the world.

447. Nye, Bill. "On Astronomy." **Life**, 42 (28 April 1910), 772.

Satire. "I sacrifice my health in order that the public may know at once of the presence of a red hot comet, fresh from the factory. No serious accidents have occurred ... not a star has waxed, not a star has waned to my knowledge." Concludes that astronomy is boring.

448. [Tessier, A.C.]. "Halley's Comet." **London Times**, 28 April 1910, 15.

Column b. Letter. Observation reports from the Cambridge Observatory; Paisley Observatory, Rye; and Southlands are given brief mention. Crommelin is quoted saying that we shall not pass through the comet's tail. The comet will be brighter than a first magnitude star.

449. Proctor, Mary. "Times Tower View of Comet." **New York Times**, 29 April 1910, 8.

Letter. Describes a probable glimpse of the comet from the top of the Times Tower. Tells would-be observers not to be disheartened by the absence of a tail.

450. "A Wonder from the East." **Sphere**, (30 April 1910), iii.

Advertisement. "Halley's Comet is supreme among celestial sights as Shem-el-Nessim is among delightful scents, but while Halley's Comet is a fleeting visitor Shem-el-Nessim has come to stay. It is indeed an inspiration in perfume." The comet soars over an Arabian motif.

451. "Progress of the Comet." **New York Times,** 30 April 1910, 2.

Column 1. Sighted at sea; no disturbance to wireless traffic has been reported. A Rutgers College astronomer reports that the comet is visible to the naked eye, though not conspicuously bright. Towaco, New Jersey chicken thieves mistaken for comet watchers went unapprehended.

452. Lynch, James K. "Halley's Comet--A Model of Its Orbit." **Scientific American,** 102 (30 April 1910), 359.

Gives instruction for the construction of a model of the comet and its orbit using paste, paper, and cardboard. Looseleaf materials referred to are not herewith included.

453. "Comet Notes [from Dominion Observatory]" **Journal of the Royal Astronomical Society of Canada,** 4 (May 1910), 224-27.

Presents a table of exposures as well as observation notes for the period of 27 April through 9 June 1910.

454. "Comet Notes." **Journal of the British Astronomical Association,** 20 (May 1910), 442-48.

Lists information received from members of the Association under headings of: 1) Brightness; 2) Length of Tail; 3) Physical appearance of nucleus and coma; and 4) Physical appearance of tail. An excellent overview of data from current observations.

455. "Halley's Comet and Meteors." **Observatory,** 33 (May 1910), 219.

Though there is no definite proof of a connection between comets and meteor-streams, it seems well to point out that the n Aquarids supposedly relating to Comet Halley are predicted for 1-7 May 1910.

456. "How Will It Strike.?" **Honey Jar,** (May 1910), rear cover.

Advertisement. The Lea-Mar Print Shop of Columbus, Ohio, publisher of the journal, advertising itself, depicts the comet plunging toward the earth. "That is the question ... How will it [printing] strike those who see it.?"

457. "Notes." **Observatory**, 33 (May 1910), 216-17.

Relates observations and estimates of the comet's brightness at Saragossa, Spain; Santiago, Chile; and Bouzareh, Algiers. From these reports it is deduced that the brighter part of the tail will not reach us at the time of transit.

458. "The Approaching Plunge of the Earth Through the Tail of Halley's Comet." **Current Opinion**, 48 (May 1910), 511-14.

Synthesizing other magazine articles, the author expresses a concern that the only fear that need be felt is that after such elaborate preparation by astronomers the comet may not do as they expect. Speculates about the effect a denser tail would have on the earth ("instant death would result"). Two diagrams and one photograph are included.

459. "The Comet." **World's Work**, 20 (May 1910), 12875.

When the sky was last ablaze with the glory of Halley's comet Africa was an unknown continent, Asia a land of mystery, Japan a hermit nation. This year the comet ought to terrify no one in the civilized world. As we watch for the phosphorescent glow in the evening sky of May 18th, we shall all be thrilled a little at the thought that we are surrounded by corpuscles which have been swept to us out of the depths of space far deeper than the flying earth ever visits.

460. ["Notes"]. **Observatory**, 33 (May 1910), 226.

Clippings from the news media: From the **Central News**, "Terror of the Comet: Prayer and Fraud in South Russia;" from the **St. James Gazette**, "in Bavaria the comet has already caused a rise in the price of beer."

461. ["Report of Meeting"] **Journal of the British Astronomical Association**, 20 (May 1910), 404-06.

A report of communications received by the Association regarding observation of Halley's Comet. Londoners were concerned--the comet had gotten itself a bad reputation. Most Londoners looked upon it as a distinct fraud, but this was the fault of geographical position and the London atmosphere.

462. Cowell, P.H. "Halley's Comet." **Contemporary Review**, 97 (May 1910), 580-88.

Of the twenty or so periodic comets only Halley's is visible to the naked eye. It alone among comets can be traced back to before the invention of the telescope.

Cowell presents his description of the comet's nature and periodicity in layman's terms .

463. Dean, John Candee. "Halley's Comet and the Church." **Popular Astronomy**, 18 (May 1910), 295-96.

The author seems to mistake the comet of 1680 for Halley's (which appeared in 1682). In essence, this article only thinks it is about Halley's Comet.

464. Eddie, L.A. "Halley's Comet." **Journal of the British Astronomical Association**, 20 (May 1910), 427-30.

Observations at Grahamstowm, England for the latter half of April 1910. The comet is conspicuous to the naked eye and is seen to have undergone a change in the period of a few days--its brightness has declined. The southern border of the tail is brighter than the northern.

465. Goatcher, A. Winton. "Notes on Halley's Comet, April 13 to May 3, 1910." **Journal of the British Astronomical Association**, 20 (May 1910), 430-33.

Describes his primitive photographing of the comet. Says that the wisps and knots and streamers remind one forcibly of Morehouse's Comet.

466. Griffin, Frank Loxley. "Halley's Comet and Others." **Chautauquan**, 58 (May 1910), 400-18.

A general historical review of the comet and historical events associated with its appearances. The last part of the article deals with cometry in general and is not specific to Halley's Comet. Photographs and diagrams of Halley's Comet and its path are provided.

467. Hawks, Ellison. "Halley's Comet in 1682." **Journal of the British Astronomical Association**, 20 (May 1910), 439.

Letter. Reference to the comet was made by Ralph Thoresby in his diary on 25 August 1682. Thoresby alludes to lingering superstition in the face of the knowledge that the comet stems from natural causes. An interesting insight into an educated mind responding during the "Age of Reason" to what would become known as Halley's Comet.

468. Hunter, A. F. "A Layman's Diary of Halley's Comet, 1910." **Journal of the Royal Astronomical Society of Canada**, 4 (May 1910), 204-11.

Observation notes for 28 April through 3 June 1910 at Barrie, Ontario. Not only is there vivid description of the comet, but also an excellent rendition of rudimentary amateur methodology employed to provide fairly accurate data--no curvature of the tail was seen.

469. Michell, S. Alfred. **Monthly Evening Sky Map**, 4 (May 1910), 4 pp.

Diagrams giving dates and position of Halley's Comet for May 1910 in the western evening sky are presented.

470. Phillips, T.E.R. "Observations of Halley's Comet." **Journal of the British Astronomical Association**, 20 (May 1910), 426-27.

Observations taken on 18, 22, 23 May. The estimated brightness on the 18th was nearly equal to that of a first-magnitude star. The night of the 18th was overcast and there was a thunderstorm.

471. See, T.J.J. "What We Know About Comets." **Pacific Monthly**, 23 (May 1910), 509-19.

States that the appearance of Halley's Comet makes this an appropriate occasion to recall to mind what is know about comets. Discusses comets in general with occasional reference to Halley's Comet. Presents a diagram of Halley's Comet with reference to the planets and two illustrations.

472. Wilczewski, Joseph. "The Real and Apparent Positions of Halley's Comet During Its Period of Greatest Importance." **Popular Astronomy**, 18 (May 1910), 257-62.

Refers to the comet as the most famous object in the history of cometary astronomy. Provides a diagram of the path of the comet superimposed on the orbit of earth; there is also a diagram of the comet's path as viewed from earth and a diagram of the comet's return and after present return. The next search for Halley's Comet will be in 1984.

473. "Comet As She Saw It in 1835." **Cleveland Plain Dealer** [Cleveland, Ohio], 1 May 1910, 4B.

A ninety-one year old woman distinctly remembers a night of falling stars in 1834, the year before the comet appeared. The comet was very beautiful but carried a sinister meaning that held people in fear. She knew that God was not yet ready to destroy the earth.

474. "Photograph of Halley's Comet Shows Sodium." **Arizona Daily Star** [Tucson, Arizona], 1 May 1910, 1.

Column 2. A telegram received at the University of Arizona states that photographs taken at the Lick Observatory reveal a brilliant line of sodium in the comet. It is the sodium that accounts for the comet's yellow color.

475. "Telescope Market Boomed by Comet." **New York Times**, 1 May 1910, 8.

Column 1. The demand for telescopes to view the comet has almost exhausted the supply in New York City; there are also big sales for binoculars. Pasteboard spyglasses are on sale for one dollar. Pawnshops are disposing of unredeemed glasses at a profit. Many comet novelties are now available on the market.

476. "Halley's Comet." **London Times**, 2 May 1910, 12.

Column d. Because the **Times** has learned that readers have had difficulty finding the comet with the naked eye or binoculars, a diagram is provided showing the position of the comet and Venus with reference to the great square of Pegasus.

477. "Think It Halley Comet." **Ohio State Journal** [Columbus, Ohio], 2 May 1910, 1.

In Akron a gas main burst and exploded. Many people thought it was Halley's Comet approaching.

478. "Comet Reported At Sea." **New York Times**, 3 May 1910, 13.

Column 2. The comet was seen by the steamer, Iroquois, at sea between New York City and Jacksonville, Florida. It was observed plainly with a glass at 4:30 A.M. Its position is about 25 degrees northeast of the planet Venus. No effect on wireless instruments was noted.

479. "Crazed By Fear of Comet's Dash, Shrieks of Death in World Afire." **Cleveland Plain Dealer** [Cleveland, Ohio], (3 May 1910, 1.

An Orrville, Ohio farmer who became mentally unbalanced claims that the world will end on Ohio primary election day (17 May), and "no one but God can save us."

480. "Flashing Across Skies 26th Time." **Ohio State Journal** [Columbus, Ohio], 3 May 1910, 7.

Once regarded as the harbinger of disaster, the comet is now merely an object of interest. The average speed of the comet is 1000 miles a minute, and when it passes between the sun and the earth it will be travelling at 1542 miles per minute.

481. "Gives a Comet Party." **New York Times**, (3 May 1910), 13.

Column 2. A Flatbush girl starts a new social fad, but fails to see the comet. Party numbered 13 thus defying superstition, all wore black and masks decorated with fake spiders and webs. New fad could provide men with an excuse to stay late at the lodge.

482. "Comet Parties Are Now All the Rage." **Wisconsin State Journal**, 4 May 1910, 1.

Cartoon. Five frames depicting the comet being observed from "any manufacturing district," at "comet parties in the fashionable neighborhoods," "in the rural district," "by the milkman," and "by the large family from their roof top."

483. "Halley's Comet." **London Times**, 4 May 1910, 13.

Column b. Relates a telegraph from Sir Robert Ball: "Halley's Comet was observed at Cambridge at 3 this morning. The stellar nucleus was between the second and third magnitude, and the tail was 20 minutes long."

484. **Rapid City Daily Journal** [Rapid City, South Dakota], 5 May 1910, [?].

Quotes a Sioux City, Iowa weather forecaster that the comet's approach has not affected weather conditions. Predicts that when the earth passes through the comet's tail affects will be seen all over the globe—similar to the eruption of the volcano, Krakatoa, in 1883.

485. "Halley's Comet." **London Times**, 5 May 1910, 13.

Column c. Reports on the Alfred Lecture delivered yesterday by H.H. Turner. Recent observations have indicated that the tail of the comet is too short to envelop the earth on 18 May. Notes Turner's remarking on the coincidence of results of the 1835 and the 1910 Parliamentary elections.

486. "Halley's Comet." **Nature**, 83 (5 May 1910), 290-91.

Recounts attempts to observe the comet with the naked eye or binoculars.

487. "Says Earth In No Danger." **New York Times**, 5 May 1910, 6.

Professor Jacoby, of Columbia University, spoke to an overflow crowd telling them that the earth is in no danger from collision with Halley's Comet.

488. Irwin, Wallace. "An Interview with Halley's Comet." **Life**, 42 (5 May 1910), 827 ff.

Satire. An interviewer travels to the comet which

appears in the heavens as a smear of chalk would appear on a dirty slate. The comet says it is nice to be back before Teddy Roosevelt (who is in Europe on tour). The comet recognizes the U.S. as the place were the Tariff is worshipped and the divorce code is taught in the schools.

489. Leach, C. "Observations of Halley's Comet." **Nature,** 83 (5 May 1910), 277.

Observed through field glasses 24 April 1910 at 3:40 A.M. at Malta. The comet was seen best at at 4:15 A.M., when the head could be distinguished with the naked eye for just a minute. Viewed again on the 26th from 3:45 to 4:30 A.M.

490. "Comet Is Seen by Many." **Ohio State Journal** [Columbus, Ohio], 6 May 1910, 12.

At 3 o'clock in the morning both head and tail of the comet were visible well above the horizon in the east--the head was to the north with the tail streaming off to the south.

491. "Comet Seen from Ferries." **New York Times,** 6 May 1910, 6.

Column 3. The comet was sighted above New York City in early dawn. Early morning commuters got a good view of the visitor. It was at its brightest soon after 3:00 A.M. The tail could be seen sweeping away to the southeast.

492. "Early Morning View of Halley's Comet." **Wichita Daily Times** [Wichita Falls, Texas], 6 May 1910, 2.

Presents a diagram prepared by Professor Henry Norris Russell, of Princeton University, showing Halley's Comet as it will appear over New York City on 2 May at 5:00 A.M.

493. "Light, Silvery Gowns." **Indianapolis News** [Indianapolis, Indiana], 6 May 1910, 8.

In Muncie, Indiana comet parties have become popular this week. Gowns of a diaphanous material, but veil-like, enveloping a fairy form must be worn on roof-top at 2:30 A.M.--party-goers are to come dressed like a comet.

494. "This the Way to Conduct a Comet Party." **Indianapolis News** [Indianapolis, Indiana], 6 May 1910, 8.

The latest social event in Columbus, Indiana is the comet party. Gives sample invitation (to include star dust and tail hairs). Most interesting comet parties

will be those with only one guest, although many
parents object to Reginald remaining that late.

495. "Concerning the Comet." **Collier's National Weekly,** 45
(7 May 1910), 13.

Provides a diagram of the orbits of the comet and
earth. Halley's Comet is referred to as a celestial
searchlight. The chance of any comet passing the earth
so that its rays will be focused on the earth is about
one in a billion. Some astronomers still think that
the comet of May 1909 was Halley's.

496. "Halley's Comet Seen By Many." **Morning News**
[Wilmington, Delaware], 7 May 1910, 1.

It is being watched by hundreds of Wilmington residents
who are willing to give up a few hours sleep. It is
visible for a short time before sunrise. Local
landmarks are given as observation bearings. The comet
is travelling at 1750 miles per minute.

497. Proctor, Mary. "The Comet From the Times Tower." **New
York Times,** 7 May 1910, 7.

Column 4. Includes a sketch and an illustration. The
clear night allows Miss Proctor to make a close study
of the comet. It was seen at 3:00 A.M. with an opera
glass. By 4:00 A.M. it was no longer visible.

498. Russell, Henry Norris. "The Heavens in May."
Scientific American, 102 (7 May 1910), 378.

A diagram of the night sky for April and May. The
description of astronomical phenomena is dominated by
Halley's Comet.

499. "Comet Hidden Yesterday." **New York Times,** 8 May 1910,
II, 20.

Column 2. Only its nucleus could be seen through cloud
banks. In Woodbury, New Jersey residents desiring to
see the comet can leave wake up call requests with the
police department.

500. "How You May See Halley's Comet." **Mobile Register**
[Mobile, Alabama], 8 May 1910, 8.

Column 2. Relates how pajama clad residents of Mobile
arise at 3:30 A.M. only to gaze enraptured at the
planet Venus. Venus is located almost exactly where
astronomers have declared Halley's Comet to rise.

501. Mayer, Hy. **New York Times,** 8 May 1910, V, 16.

Cartoon. A man is knocked sprawling as an angry

soaring comet looks down. Caption reads: "Here, why don't you blow your horn!"

502. Proctor, Mary. "Fears of the Comet Are Foolish and Unfounded." **New York Times**, 8 May 1910, V, 7.

Reviews how the approach of various comets in the past have caused alarm. The author has heard accounts of suicide and insanity in the face of this comet which is purported to make the ocean boil and scorch the earth. She points out that similar scares in the past all have proved to be unfounded.

503. "Comet Scare in Bermuda." **New York Times**, 9 May 1910, 4.

Column 2. Blacks in Bermuda report that the comet is acting strangely after the death of King Edward—a red tinge has been noted in the tail, as well as a sudden flaring at end of the tail, and the head glowed a red ball of fire for 15 minutes. Workers on the docks fell to their knees in prayer. See also entry 506.

504. "Comet's Tail." **Daily Picayune** [New Orleans, Louisiana], 10 May 1910, 6.

Columns 1-2. Dr. A.B. Dinwiddie of Tulane University in addressing the Louisiana Engineering Society, assures that although the earth may graze the tail, we are in no danger of being killed on 18 May.

505. "Halley's Comet." **London Times**, 10 May 1910, 13.

Column f. Reports on a talk by A.C.D. Crommelin at the Victoria Institute (see entry 300). He considers the association of the comet with historical events merely a memory aid. Crommelin speculates as to the reference to Halley's Comet in the prophecies of Jeremiah (verses 11-14) in the Bible.

506. Proctor, Mary. "Barnard Pictures of Halley's Comet." **New York Times**, 10 May 1910, 7.

Column 2. Two photographs are included. Explains the Bermuda scare (see entry 503) in terms of the light of the comet and Venus filtering through the mist of the horizon.

507. "Blame the Comet for It." **New York Times**, 11 May 1910, 5.

Column 3. The passengers and crew aboard the just arrived ship, Kaiser Wilhelm der Grosse, attributed a storm at sea, bad dreams, and a melancholy cat to the Comet Halley.

508. "Comet Is Off Its Orbit?" **Minneapolis Morning Tribune** [Minneapolis, Minnesota], 11 May 1910, 1.

Reports that French astronomers have noted that the comet's tail has increased in length from 5 to 10 degrees in the last three days; cyanogen has again been noted in a spectrograph of the comet. They think it not impossible that comet could after all affect earth's atmosphere.

509. "Comet Was Jerusalem's Omen." **New York Times**, 11 May 1910, 5.

Column 3. A.C.D. Crommelin is reported as stating in a lecture (see entry 300) that the sword Josephus said hung over Jerusalem before its fall was probably Halley's Comet.

510. "Comet's Tail Poisonous." **New York Times**, 11 May 1910, 5.

Column 3. An astronomer at the Paris Observatory thinks the comet's gases might affect earth's atmosphere. Other French astronomers find an important variation from the comet's predicted orbit that might bring unexpected results.

511. "Gets Good View of Comet." **New York Times**, 11 May 1910, 5.

Column 3. Dr. William R. Brooks reports excellent observation of the comet at Hobart College in Geneva, New York. The nucleus is brighter than any star in that region of the sky.

512. "Halley's Comet." **Daily Picayune** [New Orleans, Louisiana], 11 May 1910, 8.

Column 3. The comet is well above the horizon by 3:30 A.M. and is rapidly developing the tail associated with comets. It is being closely watched all over the world and, because of advances in the scientific equipment at astronomers disposal, we shall know more about it than ever before.

513. Nankive, Frank. "Watching the Comet." **Puck**, 67 (11 May 1910), [10].

Cartoon. Caption reads, "Once every seventy-five years the roof is mightier than the Parlor." Depicts a man and a woman cuddling on a rooftop ("in year thirty-five") as the comet soars above.

514. Proctor, Mary. "Brief View of Comet." **New York Times**, 11 May 1910, 5.

Column 3. A brief description of the view from the Times Tower. The comet remained visible to naked eye for 10 minutes on the morning of the 10th. The tail is now fan-like in appearance and was almost 5 degrees in length. The city's lights cut down visibility.

515. **Nation,** 90 (12 May 1910), 473.

The only persons subject to the terror of this comet are men like the hero of Branson Howard's play, who says, "I remember the prayers my mother taught me, and I say them every morning before I go to bed."

516. "Comet May Darken Air." **New York Times,** 12 May 1910, 6.

Column 3. Dr. R.W. Willson says the comet may darken the earth's atmosphere and cause electric particles to glow as the earth passes through the tail of Halley's Comet.

517. "Facts About Halley's Comet." **Journal of Education,** 71 (12 May 1910), 523.

Selected facts: Velocity=40 miles per second; head over 200,000 miles in diameter; 30 million miles long; nearest to earth at 14 million miles away; brightest from earth on 25 May; best observed from Pacific islands; best seen in morning sky 10-16 May, then in evening for remainder of May.

518. "Halley's Comet and Meteorology." **Nature,** 83 (12 May 1910), 320-22.

An account of observations from Milan, Malta, Cape Town, various parts of England, with a diagram of the comet as seen on 2 May 1910 at 4:00 A.M. at Aquila.

519. Campbell, Frederic. "Thru the Tail of Halley's Comet." **Independent,** 68 (12 May 1910), 1012-15.

A diagram of the path and the scheduled times of Halley's Comet is provided. If the visit of the comet is an event of a lifetime, then the passage of earth through its tail is an event of the ages. There is no danger of a collision--we are safe as long as there is a God.

520. "Comet Scares French." **New York Times,** 13 May 1910, 1.

Column 2. Unseasonably bad weather in France is being attributed to the comet. Even many of the cultured classes fear the comet.

521. "Got Pictures of Comet." **New York Times,** 13 May 1910,

1.

    Column 2. Balloonists A.H. Forbes and J.C. Yates say that they made photographs of the comet from 18,000 feet aloft at Horse Cave, Kentucky. Returning to earth they escaped unharmed as they fell the last 300 feet.

522. "Halley's Comet." **Baylor County Banner** [Seymour, Texas], 13 May 1910, 3.

    Those who had occasion to be awake at 3:00 A.M. on to daybreak this week have been rewarded by a sight of Halley's Comet. It is a brilliant affair and has a tail 2 million miles long. In ancient times the visitor was greatly feared as an evil omen.

523. "Mobile Negroes Alarmed Because of Halley's Comet." **Mobile Register** [Mobile, Alabama], 13 May 1910, 3.

    Columns 1-2. The last three days have offered Mobilians a splendid view of the heaven's most rare and beautiful attraction. It might be truthfully said that more colored persons are watching the comet than are white persons. The report goes on to relate the superstitious reaction of the black population to the comet.

524. Astronomer Royal [W.H.M. Christie]. "Observations of Halley's Comet Made with the Altazimuth of the Royal Observatory, Greenwich." **Monthly Notices of the Royal Astronomical Society**, 70 (13 May 1910), 539.

    Observations made 2, 6, 7, and 9 May 1910. Assumed mean places of stars are given. On 6 May the nucleus of the comet appeared slightly oval, about 3 inches in diameter, equivalent to 800 miles.

525. "Bird's Custard." **Sphere**, (14 May 1910), iv.

    Advertisement. A cup of Bird's custard is the nucleus of a comet, on the tail of which is lettered "Bird's Custard." The comet soars toward the earth. The heavens state, "Attracts the World."

526. "Children See the Comet." **New York Times**, 14 May 1910, 7.

    A teacher takes public school classes to the Park Observatory at 2:30 A.M. The mothers who brought the children to the meeting place did not wish to continue on to view the comet.

527. "Comet View from Balloon." **New York Times**, 14 May 1910, 7.

    Column 3. Roswell C. Tripp, a New York stockbroker,

has departed for Springfield, Massachusetts, from where he and some friends will observe the comet in a balloon from a height of at least 10,000 feet.

528. "Halley's Comet As Seen from Times Tower Yesterday Morning." **New York Times**, 14 May 1910, 7.

Columns 2-3. Presents a diagram of the comet over New York City. It was visible from 3:10 to 4:20 A.M. with a tail stretching about 25 million miles.

529. "Halley's Comet and the Talmud." **Scientific American Supplement**, 69 (14 May 1910), 320.

Relates Talmudic passage in which Joshua and Gamaliel converse about the star that appears every 70 years and deceives mariners. Ancient Hebrew text numbers are given in rounded form therefore explaining discrepancy between 70 and the correct 76.9 years. See also entry 159.

530. "Halley's Comet." **London Times**, 14 May 1910, 10.

Column c. The International Commission for Scientific Aeronautics has arranged a series of ascents for a registering balloon on 18, 19, and 20 May at 7 A.M. and also at 10 P.M. on the 18th. The balloons are to be sent up free, therefore their recovery is matter of chance.

531. "Halley's Comet." **Ohio State Journal** [Columbus, Ohio], 14 May 1910, [2?].

Reports the current astronomical wisdom pertaining to the comet. Points out that there is no need to fear danger from the comet, although it has come 21 million miles closer to the earth in the one-week period from 10 to 18 May.

532. "Involuntary Star Gazers." **New York Times**, 14 May 1910, 7.

Column 2. Mount Holly, New Jersey's newly installed fire whistle was sounded at 3:25 AM for a boat house fire. All who rose to view the fire got an unexpected glimpse of Halley's Comet on their way home to bed.

533. "Mitchell Ridicules Fear of the Comet." **New York Times**, 14 May 1910, 7.

Column 4. A Columbia University professor tells members of the Thirteen Club that the comet has aroused superstition for ages. He attributes allusion to the comet to Homer in the **Iliad** and to Shakespeare. He also points out that the earth has already passed through comets unscathed.

117

534. "Sea Telephones Affected." **New York Times**, 14 May 1910, 7.

The comet was observed at 3:15 A.M. from the ship, Antilles. A continued loud ticking was heard through the phones attached to the ship's wireless, and many shooting stars were observed to the southeast. The length of the comet's tail was at least 40 degrees.

535. "Tampans Called to Greet Comet." **Tampa Morning Tribune** [Tampa, Florida], 14 May 1910, 2.

Column 2. The newspaper has called more than 150 citizens by telephone to see Halley's Comet. Yesterday morning the heavenly visitor shown with great brilliance. Wakeup calls monopolized the local telephone exchange.

536. "Use Hotel Roofs for Comet Parties." **New York Times**, 14 May 1910, 7.

Roofs of all the big hotels are crowded nightly with comet parties. Clerks are kept busy calling guests who wish get up and view comet. Amateur astronomers had a 2:30 A.M. breakfast at the Gotham Hotel in an improvised bower while an orchestra played.

537. "Will It Strike?" **Daily Picayune** [New Orleans, Louisiana], 14 May 1910, 1.

Cartoon. Reprinted from the **Nashville Banner**. The earth labelled "Administration" with a perplexed elephant and politician sitting atop it peer off into space at an approaching comet (labelled "trouble") with its teeth bared.

538. Humphreys, W.J. "Atmospheric Phenomena and Halley's Comet." **Bulletin of Mount Weather Observatory**, 2 (14 May 1910), 279-85.

Comments on the potential astrophysical and meteorological effects of the comet: gross appearances, spectrum, polarization, light fluctuation; electrical potential, atmospheric conductivity, damping of electrical waves, earth currents, auroral displays, etc.

539. Jacoby, Harold. "The Case for the Earth." **Harper's Weekly**, 54 (14 May 1910), 12, 30.

Rebuts those who see a danger to the earth in Halley's Comet. The author asserts that we may greet the comet with the confident expectation of an interesting celestial event. Presents some interesting basic statistics about the comet. This is half of a pro/con pair of articles on whether or not the comet holds any danger for the earth. See also entry 540.

540. McAdam, D.J. "Will Halley's Comet Harm the Earth." **Harper's Weekly**, 54 (14 May 1910), 11-12.

A diagram of the comet's path around sun, showing its position on 18 May is provided. McAdam points out that there are many comets, and for the most part, they do not arouse concern. But Halley's Comet has history, and it strikes terror in people. Popular astronomy and history of the the comet is related. This is half of a pro/con pair of articles about whether the comet holds any danger for the earth. See also entry 539.

541. Naulty, Edwin Fairfax. "The Transit of Halley's Comet." **Collier's National Weekly**, 45 (14 May 1910), 46.

During the time that the comet is transiting the sun's disk, the earth will be swept by the celestial search light. There will be no danger from the tail, but there might be from the focal point as it sweeps across rays of sunlight which it condenses.

542. Rehn, Frederick J. "Venus Mistaken for the Comet." **New York Times**, 14 May 1910, 7.

Column 5. Letter. Rehn details his observation of the comet in from the Catskills. An editor's note states that what Rehn observed was really Venus. One wonders why editor saw fit to publish this letter unless he thought it generally instructive to deride such preoccupation with the comet.

543. Russell, Henry Norris. ""Halley's Comet In the Evening Sky." **Scientific American**, 102 (14 May 1910), 394.

Provides a diagram: The track of Halley's Comet and neighboring stars, 20-30 May 1910. The average man will find the last ten days of May the best opportunity to observe the comet.

544. "Comet Exhibited at American Museum." **New York Times**, 15 May 1910, 2.

Column 3. To show how Halley's Comet looks in relation to Mars, Venus, and other celestial bodies, an ingenious apparatus has been set up in the museum's foyer. A small model of the comet is fixed to a rod and changed as the comet changes its position.

545. "Comet Recorded in 1066." **New York Times**, 15 May 1910, 2.

Column 1. A "Regulae Canonoricum" of the year 1066 has just been found in the archives of the Cathedral of San Lorenzo in Viterbo, Italy. Halley's Comet is recorded as appearing on 5 April and flaming until 24 April.

546. "French Are Much Disturbed." **New York Times**, 15 May 1910, 2.

Column 1. Unseasonably bad weather has been blamed on the comet by Parisians. In spite of a superficial air of indifference and skepticism, the menace of the comet is really taken rather seriously.

547. "Halley's Comet Composition and Structure." **Arizona Daily Star** [Tucson, Arizona], 15 May 1910, 3.

Column 1-2. Reports observations made by the 60-inch reflecting telescope at Mt. Wilson. Describes what will happen during the earth's transit through the comet's tail, striving to allay any apprehension the public might have.

548. "Halley's Comet in the Bayeux Tapestry, A.D. 1066." **New York Times**, 15 May 1910, 2.

Columns 2-3. Presents a reproduction of that part of the tapestry that depicts the comet terrifying the Saxon king, Harold, while inspiring the Normans on the eve of their invasion of England.

549. "Heads or Tails." **Daily Picayune** [New Orleans, Louisiana], 15 May 1910, 15.

Cartoon. The earth is portrayed as President Taft with the streaking comet's head personified by Theodore Roosevelt.

550. "Played Hookey to See Comet." **New York Times**, 15 May 1910, 2.

Column 3. A twelve-year-old boy was presumed lost after being away from home for 24 hours. Armed with four sandwiches and a dime novel he had secreted himself in the tower of his school overnight in order to view the comet, while search parties scoured the woods for him.

551. "The Chief Attraction of the Coffee World." **Los Angeles Times Weekly Illustrated Magazine**, 15 May 1910, 640.

Advertisement. Newmark's pure high grade coffee is the nucleus of a comet soaring over the earth. The comet's tail reads "leaves a delightful lingering aftertaste."

552. "Watching the Comet All Over the World." **New York Times**, 15 May 1910, 1.

Column 7. Professor Hall of Oxford University predicts

that the tail is too short to reach the earth, while American estimates make it long enough with some to spare.

553. "Will Again View the Comet They Saw 75 Years Ago." **New York Times,** 15 May 1910, V, 2.

Profiles four who saw the 1835 apparition. More about the people than about the comet.

554. Dinwiddie, A.B. "Comets." **Daily Picayune** [New Orleans, Louisiana], 15 May 1910, 8.

Columns 1-7. Diagrams of comet's orbit in solar system and in conjunction with earth during May 1910 are provided. The article presents some astronomical history of the comet and explains the current apparition. The author is astronomer at Tulane University.

555. Perplexed. "Burning Daylight." **New York Times,** 15 May 1910, II, 10.

Column 4. Letter. The author complains about the fears of his maid and wife in regard to the comet and his daughter's indifference.

556. Proctor, Mary. "Clouds Dimmed the Comet." **New York Times,** 15 May 1910, 2.

Columns 2-3 . This observer got only a brief glimpse of the comet yesterday from the Times Tower. Owing to haze and light and drifting clouds, only the nucleus and about 2 degrees of the train could be seen.

557. "Calls Comet Searchlight." **New York Times,** 16 May 1910, 2.

Column 5. Edwin F. Naulty says comet's tail is sunlight condensed from radiant solar energy.

558. "Comet to Sweep Earth this Week." **Topeka Daily Capital** [Kansas], 16 May 1910, 1.

Column 5. Various reports from astronomers around the country reaffirm that there will be no harmful effect experienced when the world passes through the comet's tail, but the earth's inhabitants may see heavenly fireworks--though scientists are not certain to what extent or in what manner.

559. "Earth's Passage Through the Comet's Tail." **New York Times,** 16 May 1910, 2.

Columns 2-3. A diagram of the earth's passage through the tail beginning Wednesday night, 18 May, at 11:20

P.M. and ending Thursday morning, 19 May, at 1:20 A.M. EST.

560. "Halley's Comet Is Fast Plunging Earthward." **Montgomery Journal** [Montgomery, Alabama], 16 May 1910, 1.

Columns 5-6. Montgomery people who are interested in the comet are breathlessly awaiting the comet's 18 May transit across the face of the sun. Speculates as to what would happen if a comet hit the moon. Thankful that things are manifestly arranged for the earth's survival.

561. "In Comet's Tail on Wednesday." **New York Times**, 16 May 1910, 1.
Column 7. European and American astronomers agree that earth will not suffer from passing through the comet's tail. Estimates are that the tail will be 20 to 40 million miles long.

562. "More Astronomers' Views." **New York Times**, 16 May 1910, 2.

Column 5. Gives differing views of New England astronomers. Some expect luminous display, while others foresee no effect at all.

563. "Tail of Heavenly Rover Won't Worry Old Earth." **Minneapolis Morning Tribune** [Minneapolis, Minnesota], 16 May 1910, 1.

"Halley's dragon" will be nearest to the earth on Wednesday when the earth passes through the tail. London astronomers will not predict what sensations will be felt then. However, New England astronomers agree that the comet is harmless.

564. "The Fear of the Comet." **New York Times**, 16 May 1910, 8.

Column 6. Correspondent supplies an early Oliver Wendell Holmes poem, "The Comet." The poem supposedly depicts the apprehension that was felt on the approach of Halley's Comet in 1835. The poem is reprinted in its entirety. However, the correspondent is in error. This poem could not have been inspired by Halley's Comet, even though it is published in Holmes's **Poems** (1836), because it was first published in **New England Magazine** (April 1832).

565. "The Tail of the Comet Is Not Composed of Gas." **Houston Post** [Houston, Texas], 16 May 1910, 1.

Edwin F. Naulty of New York has asserted that the tail of Halley's Comet is not composed of gas--but based on his own theory, that rather it is sunlight condensed from radiant solar energy. There will be no contact of the tail with the earth, nor will there be any poisonous gases.

566. Barnard, E.E. "Observations from Balloons." **New York Times**, 16 May 1910, 2.

Column 3. Arrangements have been made in Milwaukee and St. Louis for observation of the comet from balloons should the weather be such that ground observation is not possible.

567. Brashear, J.A. "No Physical Effect Likely." **New York Times**, 16 May 1910, 2.

Column 4. There is no danger to the earth. The author has observed stars through 200,000 miles of the tail without any apparent diminuition of the light of any of the stars.

568. Campbell, W.W. "Our Atmosphere a Shield." **New York Times**, 16 May 1910, 2.

Columns 2-3. The passage of the earth through the tail of Halley's Comet will have no perceptible effect upon terrestrial life, animal or vegetable. However, it is possible that the electronic condition of the atmosphere will be appreciably affected.

569. Crommelin, A.C.D. "Comet May Cause Aurora." **New York Times**, 16 May 1910, 2.

Column 1. It is hardly likely that the tail can contain any meteors. There is no probability that any sensible effect will be produced on our atmosphere. Aurora may be produced by excitement of some rare gases in the upper air stream by electrons ejected by Sun.

570. Giacobin, Michael. "Comet Is No Menace." **New York Times**, 16 May 1910, 2.

Column 1. The author, from the Paris Observatory, regards the comet as a magnificent spectacle which should be enjoyed as such.

571. Gill, David. "Thinks Tail Is Too Short." **New York Times**, 16 May 1910, 2.

Column 2. The president of the Royal Astronomical Society thinks that it is very doubtful that the comet's tail will be long enough to reach the earth.

572. Mitchell, S.A. "Earth in No Danger." **New York Times**, 16 May 1910, 2.

Column 4. Mitchell, a Columbia University professor,

explains why we are safe from the comet. The rarity of matter in the comet's tail, the scarcity of cyanogen there, and the density of the earth's atmosphere in comparison to the comet make it absolutely certain that no harm will come to us.

573. Proctor, Mary. "The Comet Disappears." **New York Times**, 16 May 1910, 2.

Columns 4-5. Clouds obscured its farewell visit yesterday. Proctor recounts the current belief that if some air is bottled on 18 May during the comet's transit, some of the comet may then be able to be passed down to future generations.

574. Rudaux, Lucien. "Too Tenuous to Enter Air." **New York Times**, 16 May 1910, 2.

Columns 1-2. The precise nature of the tail is still not known, however spectroscopic examination has shown it to contain hydrocarbons, oxide of carbon, cyanogen, and azote. The rarified nature of these gases precludes Halley's Comet from entering the earth's atmosphere.

575. Schaeberle, J.M. "Comet Theories Speculative." **New York Times**, 16 May 1910, 2.

Column 4. Schaeberle sees the present data as insufficient to determine the effect on the earth of the visit of Halley's Comet.

576. **Daily Sentinel** [Nacogdoches, Texas], 17 May 1910, 2.

Column 1. Astronomers assure us that Halley's Comet is entirely harmless, and will get no closer than 45 miles with its tail. That is some comfort, but at about 2:00 A.M. the searchlight and skywizzer seemed right in the backyard.

577. "Anxieties With No Basis." **New York Times**, 17 May 1910, 8.

Column 4. Editorial. While it is true that whatever the most learned astronomers know about comets is considerably less than they still have to discover, all agree that there is no danger from Halley's Comet. Fear of comet is similar to Millerite predictions of the world's end during the mid-nineteenth century.

578. "Comet Crosses the Sun." **Aspen Democrat Times** [Aspen, Colorado], 17 May 1910, 1.

There is "intense interest evoked by the rare and scientifically important phenomenon in the sphere of astronomy." The interest of the unscientific masses is not altogether free from the fear of possible consequences of the coming event.

579. "Comet Fascinates Paris." **New York Times**, 17 May 1910, 1-2.

Column 3. Special suppers are to be held, comet postcards are being sold, and brooches and other comet memorabilia available.

580. "Comet Scare in a Car." **New York Times**, 17 May 1910, 1.

Column 2. Passengers in an auto wreck thought they were side swiped by the comet. The accident was really caused by a fan falling five stories on to the car.

581. "Comet Watches Planned." **New York Times**, 17 May 1910, 2.

Column 3. Hotels prepare for guests who will await the passing comet's tail. It is predicted that thousands will take part, and that the excitement will rival that of watching in the New Year.

582. "Comet's Tail Only Sunlight." **New York Times**, 17 May 1910, 2.

Column 4. Letter. If the tail was dense enough to reflect sunlight, as was suggested in the **Times** (see entry 557), should not the sun's refracted light thus lag at the close of the comet's journey of 12 million miles at a right angle to the path of the scorching head?

583. "Due in Comet's Tail Tomorrow, 10:15 P.M." **New York Times**, 17 May 1910, 1.

Column 3. Halley's Comet will be in transit for 1 to 7 hours. There will be an unusual glow in the heavens and possibly a few falling stars as the earth passes through the comet's 24 million mile long tail on 18 May.

584. "Earth and Comet Cannot Collide." **Tampa Morning Tribune** [Tampa, Florida], 17 May 1910, 10.

Column 2. A collision with the ghostly visitor is now an utter impossibility. This should be welcome news to the considerable number of Tampans who have felt fear about the comet.

585. "Fire Alarm for the Comet." **New York Times**, 17 May 1910, 2.

Column 4. In Boston, Mayor Fitzgerald plans to let

those interested in viewing the comet get some sleep and then sound the city's fire alarms at around 3 o'clock when the comet is first sighted by the Harvard Observatory. Half the populace is estimated to be interested.

586. "Gets Rich on Comet Pills." **New York Times**, 17 May 1910, 2.

Columns 3-4. A Haitian voodoo man is busy dispensing sugar pills that guarantee safety from the comet. No native Haitian would be without the pills if he or she can afford them.

587. "Halley's Harmless Comet Itinerary Mapped Out." **Daily Picayune** [New Orleans, Louisiana], 17 May 1910, 12.

Columns 6-7. Father Anton Kunkel, astronomer at Loyola College, gave a lecture before an interested throng on the Loyola campus. The lecture was intended to inform and assure public that the comet was no more harmful the the "Bogie Man."

588. "In Tail of Comet Tomorrow Night." **Topeka Daily Capital** [Topeka, Kansas], 17 May 1910, 1.

Column 4. For two minutes less than an hour the earth will be on one side of the comet and the sun on the other. A shadow will be cast on other side of the earth, but a luminous streak will probably be visible, the astronomer at Washburn College reports.

589. "Lick Observatory Figures." **New York Times**, 17 May 1910, 1.

Column 3. The tail is 24 million miles long and 1 million miles wide; the transit will last 7 hours. It is very unfortunate that the moon, nearly full, will interfere with optical observations of the phenomena attending our passing through the comet's tail.

590. "One Political Comet Has Landed and Another Is on the Way." **Houston Post** [Houston, Texas], 17 May 1910, 1.

Cartoon. The comet is labeled "Lawler letter (clearing Ballinger)" and is colliding with the planet earth which has President Taft's face while in the distance another comet labeled "Democratic Victory" approaches.

591. "Smith Observatory Lucky." **New York Times**, 17 May 1910, 2.

Column 3. Excellent observations of Halley's Comet made. The comet's tail stretched out in broad band across the heavens to a length of more than 45 degrees. A change in comet's form has been noted.

592. "Smooth Voo Doo M.D. Selling Comet Pills." **Tampa Morning Tribune** [Tampa, Florida], 17 May 1910, 1.

Column 5. Comet pills are new to the pharmacopoeia. The negroes of Port Au Prince, Haiti know they are safe because they are well stocked with the pills. The prescription is one pill an hour up to the time the comet begins to recede from the earth.

593. D., C.S. "Halley's Comet in 1066." **New York Times**, 17 May 1910, 2.

Column 4. A quaint record of the "long-haired star" in the Saxon Chronicle (quoted) is noted. It is notable for the absence of superstition in its description of the comet.

594. Proctor, Mary. "Where Next to Look for the Comet." **New York Times**, 17 May 1910, 2.

Columns 3-4. Provides a diagram of Halley's Comet in western sky on 20 May. The comet will be visible in the evening sky after 20 May, following sunset. The appearance of the comet will be heralded by the ruddy glow of the planet Mars, which will be seen to the left of the comet.

595. **Daily Sentinel** [Nacogdoches, Texas], 18 May 1910, 2.

Column 1. Satire. Just where the comet will strike tonight is an unknown quantity, but feel sure that if it hits Texas it will be in the neighborhood of New Braunfels. Rumor has it that in other parts of the state views about the gubernatorial candidates have changed because of the comet.

596. "Anxious to View Halley's Comet." **Rapid City Daily Journal** [Rapid City, South Dakota], 18 May 1910, 1.

Column 5. The **Journal's** night watchman and a police officer had many requests to awaken people to view the comet. There is a report from Washington, D.C. explaining in astronomical terms what the comet is doing.

597. "Arrival of Comet Due This Evening." **Wichita Daily Times** [Wichita Falls, Texas], 18 May 1910, 1.

Column 1. The event will hardly be accompanied by any phenomena that will be visible here--it will be too cloudy. Quite a few citizens have been waking early during the last week in order to view the comet.

598. "Balloon Trip to View Comet." **New York Times**, 18 May 1910, 2.

Column 2. Aeronaut Harmon has invited the deans of Yale, Harvard, Cornell, Columbia, Princeton, Penn, and Virginia to make an ascent with him at Springfield, Massachusetts in order to view the comet from aloft.

599. "Beware the Comet." **Houston Post** [Houston, Texas], 18 May 1910, 1.

Halley's namesake will call on the earth tonight. Although much maligned, scientists say the visitor's tail cannot harm us.

600. "Chicago Is Terrified." **New York Times**, 18 May 1910, 2.

Column 4. Women have stopped up their doors and windows with towels in an attempt to keep the comet's poisonous gas out of their homes. Scores of women are suffering from hysteria.

601. "Children Fear Comet." **Enquirer** [Cincinnati, Ohio], 18 May 1910, 16.

School authorities have mandated that the public schools teach about Halley's Comet today in order to allay superstitions. There have been many requests for permission to be absent from school tomorrow because parents want the family to be together when the world ends.

602. "Comet Goes on Its Way, Still World Rolls On." **Oklahoma City Times** [Oklahoma], 18 May 1910, 1.

Column 1. The earth is none the worse for her near collision and is unscathed by the "deadly" gases of the comet's tail. Nobody was hurt and hundreds of over-superstitious people are thanking their lucky stars.

603. "Comet Makes Bonci Nervous." **New York Times**, 18 May 1910, 2.

Column 4. Alessandre Bonci, operatic tenor, is uneasy: "Of course, the scientists say that there is no danger, but what do they know about the mysterious laws of nature."

604. "Comet Not to Interfere with Assembly Says Com." **Daily Iowan** [Iowa City, Iowa], 18 May 1910, 1.

Column 4. "Students should not allow fear of the comet to keep them from attending assembly this morning," says the committee of the Cosmopolitan Club which has arranged the "Peace Day" program to be presented by foreign students at the University of Iowa.

605. "Comet's Coming Causes Great 'Knocking Off' of Negro Labor." **Mobile Register** [Mobile, Alabama], 18 May 1910, 1.

Columns 4-5. Dateline Atlanta, 17 May. The comet has had a most distressing effect upon labor conditions among Negroes in Atlanta, hundreds of whom positively refuse to work at all until the "comic" passes. Most are trying to get right with God.

606. "Don't Blame Comet for Rainy Weather Says U.S. Official." **Christian Science Monitor**, 18 May 1910, 1, 5.

An expert from the U.S. Weather Bureau attributes today's showers to usual causes and says the conditions are general.

607. "Falling Meteor Heralds Comet." **Cleveland Plain Dealer** [Cleveland, Ohio], 19 May 1910, 1, 9.

In Cleveland many fear disaster is at hand. The plate glass window of a downtown restaurant was blown out by high wind, whereupon a woman who had been sitting near it, asked a nearby minister to save her from the comet.

608. "Fearing Comet, Avoid School." **Cleveland Plain Dealer** [Cleveland, Ohio], 18 May 1910, 9.

Massillon, Ohio pupils were kept at home by parents frightened that the earth would be destroyed by Halley's Comet. Also, many persons attribute the poor turnout at yesterday's primary election to people's fear of the comet.

609. "First Meteoric Shower." **New York Times**, 18 May 1910, 2.

Column 4. Professor D.W. Morehouse of Drake University tonight (17 May) reported that he saw a brilliant and beautiful meteoric shower at 2:30 A.M. in the eastern sky.

610. "Halley's Comet Is Hailed Into Court." **Evansville Journal** [Evansville, Indiana], 18 May 1910, 1.

A satirical piece. Prominent citizens are called to verify comet's appearance and fail. Al Brand, an astronomer, affected to know more than Shorty Meyer who defines the difference between the comet and a star proving the comet's location by a geometrical figure of speech.

611. "Halo Around Sun Could Not Be Blamed on Comet." **Daily Picayune** [New Orleans, Louisiana], 18 May 1910, 11.

Columns 2-4. Many people mistook the halo effect around sun to be the tail of Halley's Comet--this was

129

not the case. Halo really caused by thin layer of clouds. Makes fun of the black community's superstitious response to the comet.

612. "Illumination By the Comet." **New York Times**, 18 May 1910, 2.

Column 3. Professor David Todd of Amherst College thinks there is a chance that the tail of the comet will make a picture like the northern lights.

613. "Inmates of Jackson Insane Asylum Interested in Comet." **Daily Picayune** [New Orleans, Louisiana], 18 May 1910, 13.

Columns 3-4. The superintendent of this Baton Rouge, Louisiana asylum relates the inmates' excitement over the comet, and that in Baton Rouge people all over the city are rising between 3 and 4 o'clock in the morning to catch a glimpse of it.

614. "May Be Meteoric Showers." **New York Times**, 18 May 1910, 2.

Columns 3-4. Professor Asaph Hall of the U.S. Naval Observatory doubts that there will be meteoric showers, but if there are, there will be no danger to the earth.

615. "May See Comet Today." **New York Times**, 18 May 1910, 2.

Column 3. Harvard observers think the comet may be visible in the afternoon.

616. "Movement of Comet." **Wichita Daily Times** [Wichita Falls, Texas], 18 May 1910, 2.

Despite the fact that astronomers have assured all that there is not the least danger, the inherent and instinctive superstition of the race attacks many persons, and some uneasiness has been felt in all parts.

617. "No Danger Lurks in Comet's Tail." **Enquirer** [Cincinnati, Ohio], 18 May 1910, 16.

Gives assurance that the best scientific authorities see no cause for apprehension over the comet. On the evening of the 17th after a down-pour, moonlight and rainbow gave a beautiful effect. The fearful thought this to be an advance agent of the comet.

618. "Observatory Comment on Effect of the Comet." **Christian Science Monitor**, 18 May 1910, 5.

Astronomers from observatories in the United States and

France give their comments on what they think the effects of the earth's transit through the tail of the comet will be.

619. "Reports from Observers in This and Other Lands." **Christian Science Monitor**, 18 May 1910, 5.

Presents brief reports from observatories in Paris, France; St. Louis, Missouri; Amherst and Taunton, Massachusetts; Chicago, Illinois; Detroit, Michigan; Pasadena, California; and New Haven, Connecticut.

620. "S. P. Men Don't Care About Comet." **Arizona Daily Star** [Tucson, Arizona], 18 May 1910, 7.

An anecdote relating conversations throughout Southern Pacific Railroad offices in Tucson about "tomorrow." It was presumed that this was the result of anxiety over the comet, when in reality it was pleasure over receiving tomorrow's paychecks.

621. "She Will Sit for Her Picture Tonight." **Houston Post** [Houston, Texas], 18 May 1910, 1.

Cartoon. Planet earth (Mrs. H. Comet) with motherly face cradles the swooshing bonneted comet. Caption reads: "Just a moment now, hold that expression--watch the little birdie!"

622. "Shun Work to Watch Comet." **Daily Picayune** [New Orleans, Louisiana], 18 May 1910, 3.

Column 4. The comet has had a most distressing effect upon labor conditions in Atlanta, Georgia. Blacks positively refuse to work until the comet passes on the night of 18-19 May. The comet is viewed as the light of an astral John the Baptist.

623. "Six Hours Tonight in the Comet's Tail." **New York Times**, 18 May 1910, 1-2.

Columns 7, 1. Weather may be cloudy. The earth will pass through 48 trillion cubic miles of a tail which weighs all told half an ounce! The Lick Observatory estimates for duration of passage, length of tail, and diameter of crossing are given.

624. "The Comet Is Here." **Enquirer** [Cincinnati, Ohio], 18 May 1910, 16.

After an automobile accidentally crashed into her Dayton, Ohio home, a women shrieked, "Save me! The comet is here." It was the first thing that popped into her mind.

625. "This Is the Day." **Tampa Morning Tribune** [Tampa,

Florida], 18 May 1910, 6.

Columns 2-3. Despite scientific assurance that the comet will do no harm to the earth there is nevertheless much fear of it. Reports fearful responses from capitals of the world.

626. "Transit Across Sun's Disk Will Take Sixty Minutes." **Christian Science Monitor**, 18 May 1910, 5.

As calculated by Dr. Kobold of Kiel Observatory, Germany, the time of ingress will be 3:17 A.M., 19 May, Greenwich Mean Time.

627. "Traps Laid for Dust of Comet, Due Today." **Minneapolis Morning Tribune** [Minneapolis, Minnesota], 18 May 1910, 1.

Members of the U.S. Geological Survey plan an unique experiment in which they will place glycerin coated plates on a high tower at Mt. Wilson Observatory, hoping to catch traces of cometary dust.

628. "Wireless in No Danger Says Technology Man." **Christian Science Monitor**, 18 May 1910, 5.

Reports that Halley's Comet will have a destructive effect upon wireless telegraph systems appear to be unfounded so far as Boston is concerned, according to Edmund B. Moore, ex-president of the Technology Wireless Society.

629. "Worry Over Comet Causes Suicides." **Daily Picayune** [New Orleans, Louisiana], 18 May 1910, 3.

Column 4. A well known farmer in Lawrence County, Alabama committed suicide by taking a large dose of strychnine because he thought the comet would set the world on fire. He leaves a widow and several children.

630. "Yerkes Observatory Ready." **New York Times**, 19 May 1910, 2.

Column 3. Experts and a battery of cameras and telescopes are prepared. "We will not be surprised by anything that happens Wednesday night."

631. Dunkelberger, M.S. "To Give 'The Comet'." **Indianapolis News** [Indianapolis, Indiana], 18 May 1910, 5.

Cartoon. Advertises a skit, "The Comet," to be performed at the local Y.M.C.A. See also entry 749.

632. Glackens, L.W. "The Heavenly Porter." **Puck**, 67 18 May 1910, [1].

Cartoon. On cover of the magazine. The comet faced with negroid features holds a wisk broom of light as it sweeps by a stuffy waistcoated earth. Caption reads, "Brush yo' off, suh? Ain't gwine t' be 'round ag'in foh sev'ty-five yeahs!"

633. Proctor, Mary. "Look Out for the Meteors." **New York Times**, 18 May 1910, 2.

Column 2. The author explains what to observe and how to time the meteors.

634. Williams, Gaar. "Come On With Your Comet." **Indianapolis News** [Indianapolis, Indiana], 18 May 1910, 1.

Cartoon. Shows a person representing the public ridiculously dressed in a diving helmet, umbrella, and specimen net ready for whatever might come.

635. Winters, William H. "Halley's Comet B.C. 44." **New York Times**, 18 May 1910, 2.

Column 4. The first authentic notice of what is now known as Halley's Comet appears in Seutonius' **Caesars**, where it is mentioned that just after Caesar's assassination in 44 B.C. a comet blazed for 7 days. Winters has his facts wrong, see entries 325 and 367.

636. Wisterman. "Come On You Old Comet!" **Ohio State Journal** [Columbus, Ohio], 18 May 1910, 1.

Cartoon. A man with the earth as his head has fists up saying to comet, "Go for it, Mr!"

637. **Independent**, 68 (19 May 1910), 1102.

The editors have no fear that, in spite of passing through comet's tail, this issue will reach readers. A correspondent has beaten all the astronomical records of the comet in telling that it appeared at the date of Methuselah's death.

638. "350 Astronomers Keep Vigil." **New York Times**, 19 May 1910, 2.

Column 4. Some 350 astronomers throughout the U.S. and over 1000 telescopes kept an all night vigil--this without counting amateur astronomers. There are 107 internationally registered observatories in U.S. with 318 astronomers on their staffs.

639. "Alarm on the Rand." **New York Times**, 19 May 1910, 2.

Column 3. The lower classes in South Africa are particularly anxious. It is reported that the wife of a prominent mine manager is living at the bottom of a mine shaft until after the comet passes.

640. "Anderson Saw the Tail." **Indianapolis News** [Indianapolis, Indiana], 19 May 1910, 15.

Hundreds of people in Anderson, Indiana who were in the street until after midnight decided to make a full night of it and were rewarded by seeing the tail again in the early morning. It was a delightful moonlit Spring night.

641. "Appearance of Celestial Visitor Affords a New Line of Dope for Experts." **Tampa Morning Tribune** [Tampa, Florida], 19 May 1910, 1.

Column 1. Relates the excitement caused to the superstitious around the country as a result of the the comet.

642. "Apprehension Can Now Be Dismissed." **New York Times**, 19 May 1910, 8.

Column 4. Editorial. The results of the comet's visit will probably all be good. Superstition has received another blow, general interest in astronomy has been excited, and everybody has at least learned a little something about the motion of heavenly bodies.

643. "Aurora Shines, Sun Boils As Comet's Tail Goes By." **Cleveland Plain Dealer** [Cleveland, Ohio], 19 May 1910, 1.

Column 7. Many a Clevelander fearful of the earth's passing through the comet's tail finds refuge in the city's churches. While many others, fascinated, throng the streets to better view the comet in a carnival atmosphere.

644. "Balloon Off to Explore." **New York Times**, 19 May 1910, 2.

Column 7. At St. Louis balloons were sent aloft to explore the comet's tail. Equipment included an aneroid barometer and a thermometer--a landing is planned at about midnight.

645. "Baton Rouge Negroes Rebel." **Daily Picayune** [New Orleans, Louisiana], 19 May 1910, 3.

Column 3. Many blacks in the city were opposed to working on 18 May because of fear that the comet's visit would mean the end to all things, and they wanted to be home with the old women and children when the end came.

646. "Berlin Comet Picnics." **New York Times**, 19 May 1910, 2.

Column 3. Countless comet picnics have been organized. Many thousands of persons will congregate in parks and open air restaurants. There are special steamship excursions on the lakes around Berlin starting at midnight. Comet hats for women are popular with pearl comet hatpins.

647. "Bets on World's End." **Cleveland Plain Dealer** [Cleveland, Ohio], 19 May 1910, 1.

Column 7. The president of the Cleveland stock exchange took a wager yesterday at odds of $1000 against a nickel that the comet's tail would not extinguish all the inhabitants of the earth.

648. "Comet Candle Destroys House." **Daily Picayune** [New Orleans, Louisiana], 19 May 1910, 3.

Column 3. In Morgan City, Louisiana a tenant house on a plantation was destroyed by fire as a result of candles that the Italian residents had burned to ward off the danger of the comet.

649. "Comet Facts, Figures, and Historical Coincidences." **Minneapolis Morning Tribune** [Minneapolis, Minnesota], 19 May 1910, 1.

Presented is a chart of historical events that occurred in the years that Halley's Comet has appeared, from 66 A.D. to 1682 A.D.

650. "Comet Fails to Strike Earth." **Wichita Daily Times** [Wichita Falls, Texas], 19 May 1910, 1.

A series of bulletins from observatories around the country reporting that the earth did not seem to pass through the comet's tail.

651. "Comet Finds Nation Looking Upward." **Register and Leader** [Des Moines, Iowa], 19 May 1910, 1.

Column 5. Everyone took a peek last night at "wandering willie." The atmosphere was bottled, pictures and "squints" taken, prayers said, while the old earth survived the brushing.

652. "Comet Gazers See Flashes." **New York Times,** 19 May 1910, 1-2.

Columns 7, 1-2. Describes many people out looking for the comet. The comet seems to have lost its girth compared to previous apparitions. The Chinese section of New York seems almost nonplussed by comet.

653. "Comet Is Come Comet Is Gone Old World Still Wags Merrily Along." **Albuquerque Journal** [Albuquerque, New

Mexico], 19 May 1910, 1.

The celestial vagrant is on its way back into space leaving earth dwellers little the wiser and no better or worse. New sun spots are the only unusual phenomenon to be observed. The ignorant and superstitious who feared death may rest in peace for another 75 years.

654. "Comet Is Lost and Learned Astronomers Don't Know Whether It Hit the Earth or Not." **Denver Times** [Denver, Colorado], 19 May 1910, 1.

The "sky wanderer gets behind schedule and is seen in the east when it should be in the west--scientists in lively discussion and may use want ad to locate comet that strayed away." Scientists cannot agree as to whether the comet's tail hit the earth today or not.

655. "Comet Made Voyagers Pray." **New York Times**, 19 May 1910, 2.

Columns 5-6. The comet was the one absorbing topic of conversation among the predominantly Italian steerage passengers on the just arrived liner, Germania.

656. "Comet Reconciles Them." **New York Times**, 19 May 1910, 2.

Column 7. A wife who left home on 7 May as a result of a domestic dispute, sends for her husband because it might be her last chance before world ends.

657. "Comet Skips Along Leaving His Tail." **Toledo Daily Blade** [Toledo, Ohio], 19 May 1910, 1, 6.

The comet has come and gone, or rather has come, saluted us and is on its way back into infinite ether. Only those confined to bed ignored the comet last night in Toledo--the town's eyes were asquint, mouth agape and face apucker.

658. "Comet Switched Tail Missing the Earth." **Akron Beacon Journal** [Akron, Ohio], 19 May 1910, 1.

Columns 1-2. Akronites were disappointed, no comet was seen. No unusual gases or meteoric display accompanied the visit of Halley's Comet.

659. "Comet Unseen by Columbus People." **Ohio State Journal** [Columbus, Ohio], 19 May 1910, 1.

Despite the fears of the superstitious, the sky visitor has come and gone. No auroral display is witnessed, contrary to the popular expectation of its occurrence. The Ohio State University observatory set an attendance record last night.

660. "Comet Was Hidden." **Houston Post** [Houston, Post], 19 May 1910, 11.

Thousands looked in vain last night for the visitor. The tail swept the earth, but no evidence of any disturbance was apparent.

661. "Comet's Tail Blamed for Petty Disasters." **Minneapolis Morning Tribune** [Minneapolis, Minnesota], 19 May 1910, 1.

While thousands of Minneapolitans searched for the elusive comet, no great catastrophes were reported in the city as a result of the earth's immersion in the comet's tail. People at one local hotel comet party formed themselves into a "gas club."

662. "Desert Narrow Strip." **Daily Picayune** [New Orleans, Louisiana], 19 May 1910, 3.

Column 2. A temporary general exodus took place on 18 May from the Minnesota Point, a narrow strip of land between Lake Superior and St. James Bay. Residents were afraid that the comet would create a vacuum creating a tidal wave that would submerge their homes.

663. "Dies on Seeing Comet." **Daily Picayune** [New Orleans, Louisiana], 19 May 1910, 3.

Column 3. At Society Hill, South Carolina an old black woman arising at 4:00 A.M. to see the comet saw it and then fell dead in her yard. Blacks of this section of the state held all night prayer meetings fearing the consequences of the comet's nearness to the earth.

664. "Earth Bathes in Comet's Tail; Still Here." **Rocky Mountain News** [Denver, Colorado], 19 May 1910, 1.

Thousands congregate in high places for a view of the celestial phenomenon but see nothing unusual. On monday next the moon will be in full eclipse and then Halley's marvel will suffuse the heavens.

665. "Earth Glides Safely Through Comet's Tail." **Minneapolis Morning Tribune** [Minneapolis, Minnesota], 19 May 1910, 1.

Observers report little positive phenomena at any point. Telescopes and cameras are levelled on the sky all over the country. Valuable pictures are said to have been taken at Yerkes Observatory.

666. "Earth Penetrates Comet's Tail." **Christian Science Monitor**, 19 May 1910, 1.

Harvard University astronomers today said that the
earth passed through the comet's tail without the
slightest perceptible trace of any physical occurrence.

667. "Earth Unharmed By Comet's Tail." **Tampa Morning
Tribune** [Tampa, Florida], 19 May 1910, 12.

Column 3. The world did not come to an end. Tampa is
just the same as it was yesterday. Few people
observing the sun through smoked glasses said they saw
anything. There are no peculiar meteorological
conditions reported anywhere in the U.S. today.

668. "Earth's Plunge Through Comet's Tail Object of
Worldwide Interest Marked by Absence of Special
Phenomenon." **Pueblo Chieftain** [Pueblo, Colorado], 19 May
1910, 1.

"The comet came, the comet went, and this old earth is
no worse and no better and thus far very little wiser."

669. "Explains Comet's Origin." **Rocky Mountain News**
[Denver, Colorado], 19 May 1910, 9.

Professor T.J.J. See announced the results of years of
research. "Comets are pieces of the outer shell of
ancient nebula from which our system was developed.
The inner parts of the nebula have been cleared away in
producing sun-planets and satellites."

670. "Flames Fall from Heavens, Start Comet Gazers'
Panic." **Cleveland Plain Dealer** [Cleveland, Ohio], 19 May
1910, 2.

Comet watchers in Roselle, New Jersey were thrown into
panic when a local chemist rigged a balloon with a time
fuse and stick of dynamite to explode at 1000 feet
lighting sodium that showered the earth.

671. "Fine Photographs Obtained." **New York Times**, 19 May
1910, 2.

Column 4. One photograph recently taken at Kodaikanal
Observatory in Madras, India shows the nucleus
surrounded by a parabolic envelope, while from it
radiates a number of streamers in the form of a fan.
These show evidence of strong disturbance action.

672. "Foreigners in Alarm." **New York Times**, 19 May 1910,
2.

Column 6. Many foreigners living in New York City and
vicinity became so frightened by the comet that they
fell to their knees in the streets. The Italian
populace seemed particularly agitated.

673. "Good-by Comet for 75 Years." **Topeka Daily Capital** [Topeka, Kansas], 19 May 1910, 1.

Column 6. The glory of the comet's departure will be far greater than its coming. Tomorrow night you will be better rewarded for watching. The world placidly wends it way on none the worse for wear.

674. "Great Interest in Halley's Comet." **Rapid City Daily Journal** [Rapid City, South Dakota], 19 May 1910, 1.

Column 1. Many people in Rapid City used smoked glass so they could observe the comet as it passed in front of the sun without damaging there eyes. Some said they saw comet's shadow. Included are various observatory reports from around the country.

675. "Halley's Comet." **Devine News** [Devine, Texas], 19 May 1910, 4.

We can hardly pick up a paper but that we see something concerning the beautiful comet. The fearful all see their evil and sinful hearts more clearly thus God may use this beautiful star to accomplish some good. Every one is waiting to see the comet.

676. "Halley's Comet." **Nature**, 83 (19 May 1910), 348-49.

Synopsis of observations--worldwide.

677. "Halley's Comet." **London Times**, 19 May 1910, 4.

Column c. Report about the Swiss Aero Club's balloon ascent from Lausanne. Two Geneva astronomers will take observations of the comet. Also a lengthy report of observations made at the Kodaikanal Observatory in Madras, India.

678. "Has Halley's Comet Turned Tail and Run?" **Arizona Daily Star** [Tucson, Arizona], 19 May 1910, 3.

Columns 5-6. Renditions of public concern and anxiety based on queries made at the newspaper's office. One rumor was that the comet had "turned over." Most persons, however, took the old "Presbyterian view" that if it was going to hit, it would hit anyway.

679. "Haze in London Hides the Comet." **New York Times**, 19 May 1910, 2.

Column 3. English astronomers did not think there would be much to see. Tail's curvature and varying direction made it uncertain in view of Sir Robert Ball.

680. "Hotel [?] By Halley's Comet." **Fargo Forum and Daily Republican** [Fargo, North Dakota], 19 May 1910, 3.

The comet may not be the occasion of earthly disaster,
but it has nearly disorganized one of city's largest
hotels, moving night clerks and telephone operators to
threaten to strike. Guests leaving orders for wakeup
calls leaving little time for the staff to attend to
their normal duties. Also given is the correct formula
for declining a comet party invitation.

681. "How Houston Prepared for the Comet." **Houston Post**
[Houston, Texas], 19 May 1910, 11.

A few serio-comic incidents are related about what
people thought would be their last day. "... the
unfounded fear was displayed for the most part, by
colored folk, nevertheless large numbers of white
people" examined their consciences as a precautionary
move.

682. "In White, Await World's End." **Cleveland Plain Dealer**
[Cleveland,Ohio], 19 May 1910, 1.

Column 6. At Berlin Heights near Sandusky, Ohio "150
'Holy Rollers,' foreigners and negroes residing near
the village dressed in white tonight and held weird
services, expecting the world to come to an end when
the earth passed through tail" of Halley's Comet.

683. "It Never Touched Us!" **Cleveland Plain Dealer**
[Cleveland, Ohio], 19 May 1910, 1.

Cartoon. Two children and a dog sit laughing with an
alarm clock on ground beside them as the comet recedes
in the distance.

684. "Lovers Defied Comet." **Register and Leader** [Des
Moines, Iowa], 19 May 1910, 1.

Column 6. Only one couple in Des Moines had the
courage to wed on the day the earth was to pass through
the comet's tail. They said they were not afraid of
the comet. As a matter of fact, they hadn't even
thought of it until the bailiff asked them if they were
concerned about the "comet getting them."

685. "Mexicans Pray, Then Dance." **New York Times**, 19 May
1910, 2.

Column 7. Hundreds of Mexicans from villages along the
U.S. border gathered about crucifixes erected on
hilltops to await the appearance of the comet that was
expected to destroy the world. As the hours passed
without catastrophe, dancing and feasting replaced the
ceremony.

686. "Miners Refuse to Work." **New York Times**, 19 May 1910,
2.

Column 5. In Wilkes Barre, Pennsylvania; Denver, Colorado; and Brazil, Indiana foreign born miners refuse to work. They want to be above ground when the world comes to an end.

687. "Mistook Balloon for Comet." **Indianapolis News** [Indianapolis, Indiana], 19 May 1910, 15.

A large balloon passed over Oakland City, Indiana at about the time scheduled for the arrival of Halley's Comet. People ran from there homes to watch the balloon supposing it to be the comet. The mistake was discovered after the balloon was lost from view.

688. "Night Services in Russia." **New York Times**, 19 May 1910, 2.

Column 3. In spite of reassuring newspaper articles, many people choose to pass the night praying in church.

689. "No Results at Yerkes." **New York Times**, 19 May 1910, 2.

Column 2. Aurora lights are seen but they are attributed to sun spots rather than the comet.

690. "Not Comet Gas, But--." **Indianapolis News** [Indianapolis, Indiana], 19 May 1910, 15.

Many Muncie, Indiana factories have been crippled by an exodus of workers. There were dozens of comet parties--a great excuse to remain out late. Relates a story about a foul smelling fire in an apartment basement was mistaken by tenants for comet gas.

691. "Not Even Comet Sparks." **Indianapolis News** [Indianapolis, Indiana], 19 May 1910, 15.

Many people in Columbus, Indiana were frightened by what they thought were sparks from the sun breaking up. In reality what they saw was the cottony pollen of the cottonwood tree floating to earth. An oldtimer says people were frightened more in 1835.

692. "Paris Keenly Disappointed." **Indianapolis News** [Indianapolis, Indiana], 19 May 1910, 15.

The only excitement was caused by comet gazers falling off roofs as a result of overindulgence in strong drink at comet parties.

693. "Phenomena Predicted in the Heavens Today." **Daily Iowan** [Iowa City, Iowa], 19 May 1910, 1.

Column 3. Yesterday was "comet day" around the university. In the "wee small" hours comet parties gathered and watched for the appearance of the wanderer. Many thought they saw the comet, but it was really Venus.

694. "Porto Ricans Confess."[sic] **New York Times**, 19 May 1910, 2.

Column 5. Many parade the streets of San Juan carrying candles and chanting prayers. Many workmen failed to report to the tobacco factories and plantations.

695. "Remove Lightning Rods." **Daily Picayune** [New Orleans, Louisiana], 19 May 1910, 3.

Column 2. Residents of Neenah, Wisconsin fearful that the rods would attract dangerous substances that might accompany the comet removed all lightning rods from homes and barns.

696. "Rush to Alpine Heights." **New York Times**, 19 May 1910, 2.

Column 3. Comet dances were held at Swiss hotels and a midnight balloon ascent was made by two astronomers sponsored by the Swiss Aero Club.

697. "Some Comet Statistics." **Bismarck Daily Tribune** [Bismarck, North Dakota], 19 May 1910, 2.

There was no disturbance out of the ordinary, except that a lady in Mandan went around and paid all her debts on Tuesday, expecting the end on Wednesday. A suicide was reported from Chicago and "blind pig" evidence came freely to the Attorney General's dragnet.

698. "Some Driven to Suicide." **New York Times**, 19 May 1910, 2.

Column 6. In Chicago some people developed temporary insanity or hysteria while others became suicidal on account of Halley's Comet.

699. "Southern Negroes In a Comet Frenzy." **New York Times**, 19 May 1910, 2.

Column 5. Many workers refuse to do field work and have flocked to town to attend all night church services. According to Ashville, N.C. police, crime among the Negro population has slumped in last few days.

700. "Spent the Night in Cellars." **Indianapolis News** [Indianapolis, Indiana], 19 May 1910, 15.

In Lafayette, Indiana last night hundreds of people forsook their beds and remained up to see what would happen. Many cases were reported of people going into their cellars and remaining until daybreak. Blacks were especially frightened, but also whites.

701. "Sun Spots Appear; Not Due to Comet." **New York Times**, 19 May 1910, 1.

Column 5. Sun spots were reported by St. Louis astronomers, but were probably not related to Halley's Comet. Other U.S. observatory reports are also given in regard to sun spots.

702. "Swish of Tail Fails to Harm Gay Old Earth." **Ohio State Journal** [Columbus, Ohio], 19 May 1910, 1.

The world passes through the comet's appendage while myriads of eyes are turned toward the heavens. The comet came and the comet went and this old earth is no worse and no better, and, thus far, very little wiser.

703. "The Call of the Comet." **Tampa Morning Tribune** [Tampa, Florida], 19 May 1910, 6.

Column 2. Wide-eyed persons in all parts of Tampa yesterday peered at Halley's Comet through smoked glasses. The comet was viewed as a demonstration of the Supreme Order that exists even in apparent chaos--in that such exact scheduling 75 years ahead was possible.

704. "The Comet." **Nation**, 90 (19 May 1910), 504-05.

People are searching the heavens not in terror, but in curiosity. What terror there is can be attributed to irresponsible yellow-journalism. Paris is planning comet parties--gentlemen wear pale blue and ladies gowns the color of the firmament.

705. "Thought It Was the Comet." **Indianapolis News** [Indianapolis, Indiana], 19 May 1910, 15.

Operations were temporarily suspended at a coal mine outside Evansville because of an elevator accident that partly wrecked the frame of a building. The miners were thrown into a panic as many thought Halley's Comet had struck the earth.

706. "Thousands Out But Comet's Show Fizzled." **Fort Wayne Sentinel** [Fort Wayne, Indiana], 19 May 1910, 9.

Fort Wayne folk waited for the sky spectacle that never came. Hundreds sought elevated spots and the streets were lined for several hours as people scrutinized the sky. Practical jokers are telephoning their friends urging them to come out of the cellar.

707. "Thousands Watch As Comet Passes." **Cleveland Plain Dealer** [Cleveland, Ohio], 19 May 1910, 1, 2.

Comet gazing parties will be formed in the public schools next week. Teachers will have charge of the children and in the evening will conduct them to some spot where the comet may be seen in order to explain to them what is known of it.

708. "Thunder Scares Boston." **New York Times**, 19 May 1910, 2.

Columns 6-7. A sudden boom of thunder followed immediately by a darkening of the sky and downpour gave nervous folk who were anticipating strange happenings on account of the comet a mild shock.

709. "Vie in Vain with Comet." **Cleveland Plain Dealer** [Cleveland, Ohio], 19 May 1910, 2.

In Newark, New Jersey three U.S. senators spoke to a sparse crowd of only 200 persons because most people were out viewing the comet.

710. "We Have Passed Through Tail of Halley's Comet." **Burlington Free Press** [Burlington, Vermont], 19 May 1910, 1.

"The comet came, the comet went, and this old earth is no worse and no better and, thus far, very little wiser." Comet parties were held everywhere--in streets, on roof tops, in gardens, but very little of the comet's tail was seen. Gives observation table for May.

711. "We're Through It the Comet's Tail." **Topeka Daily Capital** [Topeka, Kansas], 19 May 1910, 1.

Column 7. New York was excited by reports from Topeka that a meteorite had fallen through a window in the Shawnee Building. It was really nothing more that a schoolboy amusing himself by throwing bricks. Nobody knows how the news got to New York .

712. "Weather Balloons Aloft." **New York Times**, 19 May 1910, 2.

Column 7. The Blue Hill Observatory at Hyde Park sent up a sonder Balloon (6 inches in diameter) at 5 o'clock to record weather conditions 8 or 10 miles above earth. The plan is to have the same experiment to be made at 30 other places worldwide at exactly the same time.

713. Barnard, E.E. "Tail Appears Longer." **New York Times**, 19 May 1910, 2.

Column 7. This is the first clear morning in about two weeks at Yerkes Observatory. The tail appeared brighter than any portion of the Milky Way--it could be traced for 107 degrees with a width of 5-6 degrees.

714. Crane, William G. "It Didn't." **New York Times**, 19 May 1910, 8.

Column 6. Poem.

715. Denning, W.F. "Fireball in Sunshine." **Nature**, 83 (19 May 1910), 339.

Letter. Tells of a fireball sighted over the English midlands, computed to be at a height of 32-83 miles, travelling at 20 miles per second. It was thought by many to be Halley's Comet or a part of it.

716. Drayton, H.S. "A Planetary Scare." **New York Times**, 19 May 1910, 8.

Column 5. Letter. Recalls dire prophecies of 1881 when there was a planetary conjunction and compares it to the current reaction to Halley's Comet.

717. Mullens, E.T. "Observations of Halley's Comet and Venus." **Nature**, 83 (19 May 1910), 339.

Letter. Dated 22 April 1910. Reports that in Natal, South Africa the planet Venus was plainly visible to the naked eye at noon. The comet was visible from 4:30 to 6:00 A.M., so many natives and Europeans were gazing at Venus at midday thinking it to be Halley's Comet.

718. Wisterman. "Never Touched Us!" **Ohio State Journal** [Columbus, Ohio], 19 May 1910, 1.

Cartoon. Cigar smoking gent with earth as his head grabbing on the comet's tail and saying, "You coward! You were afraid t' hit me!" as the comet in the distance says, "Hi! Leggo my tail!"

719. "Astronomers All At Sea." **New York Times**, 20 May 1910, 2.

Column 1. Many explanations are advanced for the failure of the earth to pass through the comet's tail. One theory suggests that the comet dropped its tail which then drifted off into space. Another is that the tail's curvature kept it away.

720. "Band of Light Was the Tail." **New York Times**, 20 May 1910, 2.

Professor Asaph Hall of the U.S. Naval Observatory suggests that the comet has developed such a deep curve

in its tail, that the earth passed first through one
part of the tail, entered inside the curve, then
reentered the tail before getting clear of the comet.

721. "Character of Sun Spots." **New York Times**, 20 May
1910, 2.

Column 2. Professor Jerome S. Rickard of the
observatory at Santa Clara College has been watching
sun spots since they began to develop on 12 May. In
his view Halley's Comet may be viewed as a formative
planet, therefore it must have influence on both the
sun and the weather.

722. "Clouds Hide Comet's Tail." **Register and Leader** [Des
Moines, Iowa], 20 May 1910, 1.

Column 6. Determined to make the people of Des Moines
forget their comet craze for at least one night and get
a full schedule of much needed sleep, J. Pluvius
worked overtime yesterday afternoon to prevent all
possibility of seeing evidence of comet.

723. "Comet Crook; Venus Lures Flirt Out of Path." **Rocky
Mountain News** [Denver, Colorado], 20 May 1910, 1.

"Deceitful sneak fools all confounding scientists and
flies its tail in the eastern sky." This is not an
honorable comet. Many hysterical women spent night
holding their noses shut so as not to breathe lethal
comet gases.

724. "Comet Strikes Denver Man in Conscience; Pays Bills
Owed for Last 20 Years." **Rocky Mountain News** [Denver,
Colorado], 20 May 1910, 5.

In Greeley, Colorado there appeared a Denver man who 20
years earlier had fled in order to avoid paying debts.
Fearing that the comet would cause the world to come to
an end, he wished to face up to his sins before
judgment day.

725. "Comet's Acts Mystify; Sun Shows Apparition."
Minneapolis Morning Tribune [Minneapolis, Minnesota], 20
May 1910, 1, 3.

Supposition now holds that the rover's tail did not
swish the earth. Appearing in the eastern sky, while
observers were looking to the west for it, astronomers
at the Yerkes Observatory find themselves bewildered by
"Old Sol."

726. "Comet's Tail Swerved Missing Earth Entirely." **Carson
City Daily Appeal** [Carson City, Nevada], 20 May 1910, 1.

The earth may enter the comet's tail at any time.

Scientists are puzzled at unlooked for action of the celestial visitor. Many shooting stars were seen in the comet's tail yesterday morning after the moon had set, affording a brilliant sight.

727. "Confirms Bessell's Theory." **New York Times,** 20 May 1910, 2.

Column 4. Based on the tail lagging far behind the rest of the comet, Harvard astronomers say observations show that the tail rotates.

728. "Didn't Get Through the Comet's Tail." **New York Times,** 20 May 1910, 1-2.

Column 7, 1. Leading observatories confirm the **New York Times**'s discovery of the comet tail being in the east rather than in the west. Therefore the earth could not have passed through the comet's tail. Astronomers are at a loss as how to explain why the predicted transit through the comet's tail did not occur.

729. "Fear of Comet is Cause of Madness." **Arizona Daily Star** [Tucson, Arizona], 20 May 1910, 1.

Column 6. A woman in Anaheim, California attempted to take her own life and the lives of her children by drinking lye. She said the comet was certain to destroy the earth and wanted to spare herself and the children a fiery death.

730. "Futile Comet Expeditions." **New York Times,** 20 May 1910, 2.

Column 4. Reports that a balloon expedition at St. Louis and an observation by T.J.J. See in California were frustrated on account of fog, and there was cloudy weather in Hawaii as well.

731. "Goes Insane Over the Comet." **New York Times,** 20 May 1910, 2.

Column 5. (Blurb-one column inch). Miss Kate Van Ness of Carlton Hill, New Jersey was taken to an asylum. She became unhinged following a discussion of the the comet's appearance. She said she would follow the comet no matter where it went.

732. "Halley's Comet." **London Times,** 20 May 1910, 9.

Column d. Reports preparations that are being made for observation at Greenwich, Cambridge, Paris, Berlin, Rome, Madrid, St. Petersburg, Constantinople, Johannesburg, and Cape Town.

733. "Haze Overcast the Sky Last Night." **Grand Forks Daily Herald** [Grand Forks, North Dakota], 20 May 1910, 2.

Whether we have passed through the comet's tail seems debatable--astronomers cannot agree. The haze seemed particularly bright in the north, casting a reddish glow. The more superstitious people in Grand Forks are breathing easier now that the comet has passed.

734. "Hoadley Comet Club." **New York Times**, 20 May 1910, 2.

Column 5. Formed at the Hotel Knickerbocker during an observation party, the club will have 15 original members with one member added each year until membership reaches 75. Only relatives or heirs of the original members can be eligible for membership. It is about the only tangible thing the comet has left along Broadway.

735. "Kidding Comet Is Cause of Curious Complication for Colored Folk." **Denver Times** [Denver, Colorado], 20 May 1910, 1.

With the approach of the transit of the earth through the comet's tail many blacks bought life insurance. Since the comet's tail just missed the earth many who bought the insurance are demanding their money back.

736. "May Miss the Tail." **New York Times**, 20 May 1910, 2.

Column 3. Professor Henry N. Russell of Princeton University thinks that it is likely that we shall pass through the southern edge of the tail. The tail is curved away from the earth, therefore, although the head of the comet has crossed sun's disk, the comet's tail will reach the earth later.

737. "Passed Sun On Time." **New York Times**, 20 May 1910, 2.

Column 5. The earth is close to the nearer wall of the tail and is likely to enter it at any moment. Mt. Wilson observations have so far not revealed any special aurora lights nor any meteors.

738. "Phenomena Observed Abroad." **New York Times**, 20 May 1910, 2.

Column 5. Reports from Berlin; Bilba, Cuenca and Madrid, Spain; Lisbon; Paris; Aden; St. Thomas, West Indies; Willemstad, Curacao; and Johannesburg.

739. "Professor Leavenworth Explains Comet's Actions." **Minneapolis Morning Tribune** [Minneapolis, Minnesota], 20 May 1910, 3.

A University of Minnesota professor of astronomy offers

several plausible explanations for the comet's unexplained appearance in the eastern sky.

740. "The Best-Boomed Comet Show Still Very Much a Fizzle." **Daily Picayune** [New Orleans, Louisiana], 20 May 1910, 5.

Columns 1-2. The consensus of opinion is that Halley's Comet should be arrested for obtaining notoriety by false pretenses; everybody is saying the comet is a celestial confidence man, a sky-shooting bunco steerer, people will shake their heads in negative motion when speaking of the comet.

741. "The New Disease from Staring High Twists Victims' Necks Forever Wry." **Rocky Mountain News** [Denver, Colorado], 20 May 1910, 5.

"Oh, the comet face is a funny thing. It's eyes bulge out and its mouth's like a ring. Its tilted up, has a wonderful leer and the crick in its neck will last a year ... [says] 'have you seen the comet'." See also the accompanying cartoon, entry 746).

742. "Thinks Nucleus Not Solid." **New York Times**, 20 May 1910, 2.

Column 5. A Jesuit astronomer in the Phillipines believes that the old theory of a solid cometary nucleus is disproved. There have been excellent observing conditions from 3 Phillipine sights, and no trace of solid matter in nucleus has been revealed.

743. "Will View Comet From Balloon." **New York Times**, 20 May 1910, 2.

Column 5. (Blurb-one column inch). Professor Charles L. Doolittle of the Flower Observatory, University of Pennsylvania will make an ascension in order to observe Halley's Comet. See also entry 772.

744. ["Letter"]. **New York Times**, 20 May 1910, 8.

Column 3. Editorial. The **Times** is ambitious to maintain its place as the leading comet organ. By learning more about the comet we have, in effect, unlearned many false beliefs and superstitions.

745. Abelow, S.P. "Terror in Manhattan." **New York Times**, 20 May 1910, 8.

Column 5. Letter. Draws parallel with the comet of 1680 (not Halley's), being like a blazing sun and startling the Dutch inhabitants of New York. "But 1680 is not 1910!"

149

746. Cesare. "Lookers in Space Get 'Comet Face'." **Rocky Mountain News** [Denver, Colorado], 20 May 1910, 5.

Cartoon. See companion article, entry 741.

747. Proctor, Mary. "The Comet's Tail As Seen From the Times Tower Between 2:30 and 3:15 A.M. Yesterday." **New York Times**, 20 May 1910, 2.

Columns 2-4. Intermittent flashes varying from red, white, and yellow, were seen between 10:37 and 11:22 P.M. when a flash of unusual brilliancy startled watchers. It resembled an arch of glowing white surmounted by a crest of crimson. A diagram is provided.

748. Williams, Gaar. "Adrift." **Indianapolis News** [Indianapolis, Indiana], 20 May 1910, 1.

Cartoon. The caption is "Another Promise Broken." A forlorn little man stands atop the earth holding a staff with shirt attached as a banner reading "lost," a chart at his feet, and a rudder at his side. Planets and winds swirl by.

749. "'The Comet' Well Given." **Indianapolis News** [Indianapolis, Indiana], 21 May 1910, 5.

Review of a play, "The Comet," performed at the local Y.M.C.A. and viewed by a large crowd. The author is Simon H. Cox, a high school senior. The comet plays background to young love and disabled autos. See also entry 631 for cartoon advertising the play.

750. "A Headless Comet." **Houston Post** [Houston, Texas], 21 May 1910, 16.

The tail of the comet alone was seen from Houston. Speculates that it is possible that the tail became detached altogether or that rest of comet can still be seen in the east.

751. "Astronomer Gets No Sight of Comet During Aerial Trip." **Christian Science Monitor**, 21 May 1910, 1.

Professor David Todd of Amherst College ascended in a balloon in order to observe the comet. Taking off from Amherst, he landed near Montreal, Canada. Aloft for 12 hours, he failed to observe Halley's Comet due to the haziness of the atmosphere.

752. "Clouds Hide the Comet." **Topeka Daily Capital** [Topeka, Kansas], 21 May 1910, 1.

Column 4. Nobody knows what strange capers the astral visitor was cutting last night behind the heavy bank of rain clouds. You may still see the strange astral visitor tonight if the clouds have gone about their business.

753. "Comet Notes." **Scientific American,** 102 (21 May 1910), 416.

It is important that the chance to capture a sample of the tail of Halley's Comet was not seized. It has been suggested that the liquefaction of a large quantity of air which then could be later treated by fractional distillation to find cometary matter might have been tried.

754. "Comet in the West; Its Tail Missing?" **New York Times,** 21 May 1910, 4.

Column 1. Observers at Yerkes Observatory see only the nucleus. They speculate as to the reason for the absence of the comet's tail.

755. "Comet's Schedule Shows Increasing Visible Time Daily." **Christian Science Monitor,** 21 May 1910, 1.

Gives schedule of the comet for the remainder of May. The comet seems to be minus a tail.

756. "Confirms Times Observations." **New York Times,** 21 May 1910, 4.

Column 5. Lowell Observatory observations confirm the Times Tower observation of Halley's Comet's tail in the eastern sky at morning.

757. "Did the Earth Miss the Comet Tail by 197,000 Miles?" **New York Times,** 21 May 1910, 4.

Columns 2-4. Provides a diagram of the position of the earth when the comet's tail passed over it at 1:30 A.M. on 19 May. Reports that vestiges of the tail are still lingering in the eastern sky--there is a possibility that the earth's gravity may have detached part of the tail.

758. "Halley's Wanderer." **Northfield News** [Northfield, Minnesota], 21 May 1910, 1.

"The earth glided safely by the comet's tail Wednesday." Quotes H.C. Wilson, director of Goodsell Observatory, Carleton College, on the appearance of the comet.

759. "Lecture on the Comet." **Seattle Post-Intelligencer** [Seattle, Washington], 21 May 1910, 1.

Columns 4-6. Cartoon. Reprinted from the **Chicago**

Tribune. Married men would do well to stay home and the effect of its laughing gas will be mildly exhilarating. People who have not smiled for 75 years will bubble forth in ripping laughter.

760. "Light Line Not the Comet." **New York Times,** 21 May 1910, 4.

Column 3. Harvard astronomers think the beam of light seen yesterday was not the tail of Halley's Comet.

761. "Many Gaze at Comet." **Carson City Daily Appeal** [Carson City, Nevada], 21 May 1910, 1.

Everyone in the city went out last evening to view the comet, which was plainly visible in the western sky. The comet had the appearance of a large star covered with mist and the tail was at no time visible to the naked eye.

762. "Peculiar Gaseous Odor Was Not from Comet." **Albuquerque Journal** [Albuquerque, New Mexico], 21 May 1910, 1.

Dateline 20 May, Denver. Residents of a fashionable Denver neighborhood mistook the odor of a "polecat" chased out of the woods by dogs to be that of cyanogen gas from Halley's Comet. The mystery was solved by a postman, and now the neighborhood is laughing.

763. "Prof. Todd In a Balloon." **New York Times,** 21 May 1910, 4.

Column 5. Professor David A. Todd of Amherst College and three companions ascended at 6:43 P.M. with instruments to observe Halley's Comet.

764. "Saw Traces of the Tail." **New York Times,** 21 May 1910, 2.

Columns 2-3. W.W. Campbell of the Lick Observatory observed the tail on the morning of 20 May. It is impossible at this time to determine whether the earth passed through it or not.

765. "Seattle Crowds Seek Highest Points to Get View of the Visitor." **Seattle Post-Intelligencer** [Seattle, Washington], 21 May 1910, 1.

Column 3. So close did the comet follow the setting sun that it was impossible to secure a photograph last evening. Early evening saw the greater part of Seattle's population star-gazing. Many people were on the roofs of tall office buildings.

766. "Seen at Cincinnati." **New York Times,** 21 May 1910, 4.

Column 1. Observations are reported from Cincinnati,
Mt. Wilson in California, and Geneva, N.Y.

767. "Shaw's Star Outshines Halley's Comet." **American
Stationer,** 67 (21 May 1910), Cover.

Advertisement. "Shaw's star outshines Halley's Comet.
Higher than all others. If you want to know how high
the quality of Shaw Blank Books is ..."

768. "Ship's Compass Affected." **New York Times,** 21 May
1910, 4.

Column 5. The first electrical manifestation in
connection with Halley's Comet was reported by a
steamer captain. Compasses were thrown off by 3/4 of a
point at 8:00 A.M. on 16 May while at sea.

769. "Tail Lost in 1835." **New York Times,** 21 May 1910, 4.

Column 2. The tail of the comet is curved and the
earth did not pass through it. Although the nucleus
moves in an orbit, the tail is erratic. Just before
the comet was at its best in mid-November 1835, the
tail was unaccountably lost. Therefore there is a
precedent for the comet's current behavior.

770. "Views and Reviews." **Hondo Anvil Herald** [Hondo,
Texas], 21 May 1910, 4.

We were at first disposed to treat with levity any
suggestions of danger from Halley's Comet. But we have
changed our minds since learning of a woman who saw it
75 years ago and on seeing it again dropped dead--"she
could not comet again."

771. Proctor, Mary. "May See Comet To-night." **New York
Times,** 21 May 1910, 4.

Columns 4-5. An account of the author's observations
from the Times Tower building in New York City.

772. "Balloon Party Missed Comet." **New York Times,** 22 May
1910, 22 May 1910, II, 20.

Column 2. Clouds obscured the view of Professor
Doolittle of Flower Observatory. He reached an
altitude of 6,500 feet, but clouds were still too dense
to make any observation. See also entry 743.

773. "Like a Fan-tailed Pigeon." **New York Times,** 22 May
1910, II, 20.

Column 2. To scientists at the Carnegie Observatory at
Mt. Wilson the comet presented the appearance of a
fan-tailed pigeon. Despite a bright moon the comet's

tail was distinct.

774. "Local Comet Display Rather Disappointing." **Seattle Times** [Seattle, Washington], 22 May 1910, 10.

Column 3. "Suffusing the heavens with the 'bright' effulgence of about one candle power electric bulb and throwing downward a slanting ray of soft yellow light--so soft one needed to look only twice to see it"--it would take a great imagination to view the comet as a success.

775. "May Be a New Tail." **New York Times**, 22 May 1910, II, 20.

Column 2. Professor Frost of Yerkes Observatory propounds the theory that the comet's tail had become detached. Compares it with the temporary disappearance of the tail during Halley's Comet's 1835 apparition.

776. "New Yorkers View the Comet's Return." **New York Times**, 22 May 1910, II, 20.

Column 1. Comet watchers get a dim glimpse of the comet in the west. Only a stubby tail is seen. Astronomers believe that this may be because we are looking at it head on.

777. "Parisians Feared Comet Would Kill." **New York Times**, 22 May 1910, III, 3.

Column 1. Many persons bought supplies of oxygen to prolong their life in case poisonous gas swept the earth. Cellar refuges were prepared. Those unterrified by the comet spent the nights in the cafes--but did not exhibit their usual gaiety.

778. "Portland Folks Get Good View of Comet." **Seattle Times** [Seattle, Washington], 22 May 1910, 1, 10.

Columns 5, 3. The comet, glimmering in the west like a small candle, was viewed for an hour by hundreds of people from Council Crest in Portland, Oregon.

779. "Why We Missed the Tail." **New York Times**, 22 May 1910, 10.

Column 6. Letters offering various theories on why the earth did not pass through the tail of Halley's Comet.

780. Burleigh, M.M. ["Poem"]. **New York Times**, 22 May 1910, 10.

Column 6. Poem.

781. Mayer, Hy. **New York Times**, 22 May 1910, V, 16.

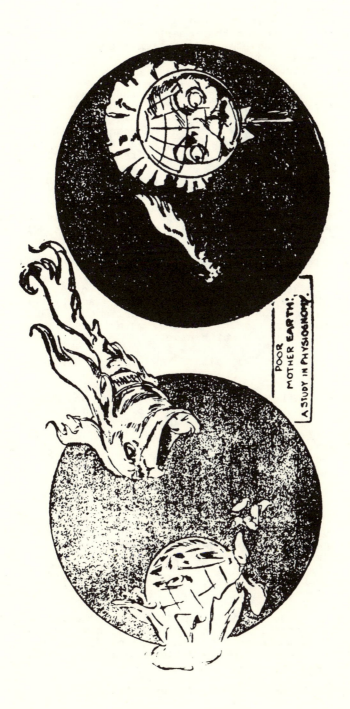

6. Cartoon by Hy Mayer, 22 May, 1910. See entry 781. © 1910 by The New York Times Company. Reprinted by permission.

Cartoon. The roaring comet approaches the frightened bonneted orb of mother earth. Next frame shows a bonneted mother earth warily tightlipped, looking at the receding comet. Caption reads: "Poor mother earth! A study in physiognomy."

782. Proctor, Mary. "Latest View of the Comet." **New York Times**, 22 May 1910, II, 20.

Columns 1-2. The new theory to account for the tail of the comet lagging behind the head is that it is being affected by Venus exerting a gravitational pull.

783. "Eclipse Will Aid Comet Sight." **Pasadena Star** [Pasadena, California], 23 May 1910, 1.

There will be a spectacular opportunity to view comet between 9:09 and 9:59. Tail is in lead of comet, but is rapidly disappearing and will soon be lost.

784. "Halley's Comet." **London Times**, 23 May 1910, 38.

Column e. Reports observations of the comet by Sir Robert Ball at Cambridge Observatory and by the many thousands of onlookers at Parliament Hill at Hampstead--"the comet was very much like any ordinary star seen through a white haze."

785. "Has the Comet Two Tails?" **New York Times**, 23 May 1910, 2.

Column 1. Photographs taken at the University of Michigan indicate that there are two tails, one a short bright one superimposed above a less distinct horizontal appendage, both extending from the sun southward. An explanation is offered for the tail being seen in both east and west.

786. "Honorary Degrees at Oxford." **London Times**, 23 May 1910, 32.

Column e. Reports the awarding of honorary degrees to P.H. Cowell and A.C.D. Crommelin for their work in investigating the movements of Halley's Comet (quotes remarks made on the occasion).

787. "Made a Comet to Order." **New York Times**, 23 May 1910, 2.

Column 1. A searchlight manufacturer mounted a 36-inch electric searchlight on the roof of his factory. In the misty atmosphere of the morning of 21 May the tail of this fake comet shooting upward made for an impressive spectacle.

788. "Meteorite or Dornick?" **Topeka Daily Capital** [Topeka,

Kansas], 23 May 1910, 1.

Column 4. Topeka resident L.B. Whitten believes that he has a real piece of Halley's Comet or at least a feather off the tail of the celestial tramp. The red hot pebble, in reality, may have come from the power plant furnace across street from Whitten's home.

789. "Negro, Crazed at Comet Murders His Own Wife." **Tampa Morning Tribune** [Tampa, Florida], 23 May 1910, 1.

Column 2. In Talladega, Alabama congregations of several churches left their pews, and hundreds of persons stood, excited, in the square and gazed at the celestial visitor (22 May). Also reported are a farm girl dropping dead on seeing comet and a Negro killing his wife and children.

790. "Observations in Finmark." **London Times**, 23 May 1910, 38.

Column e. Professor Birkeland states that on the night the comet passed the sun he observed a magnetic storm of extraordinary strength, while the nights before and after were quiet. He obtained excellent magnetograms and earth current curves.

791. "The Comet Hidden: Its Glory Dimming." **New York Times**, 23 May 1910, 2.

Column 1. The comet will only rank with third-rate stars if it is seen again. Moonlight in a golden glow behind varied clouds consoles disappointed comet watchers.

792. "Watch for Comet on Monday Night." **Montgomery Journal** [Montgomery, Alabama], 23 May 1910, 1.

Columns 5-6. Practically everyone in the city was on the lookout for the "hobo," and just after the sun went down those who were watching saw a hazy looking star which got brighter after a short time.

793. Kaempffert, Waldemar. "Instruments Used for the Study of Halley's Comet." **Seattle Times** [Seattle, Washington], 23 May 1910, 1.

Columns 1-2. A column syndicated by the Associate Literary Press in a column title the "Comet Day by Day." Describes the hardware pointed at the comet.

794. Turner, H.H. "Halley's Comet." **London Times**, 23 May 1910, 38.

Column d. Letter. There is such widespread interest in Halley's Comet that the correspondent wishes to communicate observation news that he has received from Knox Shaw at Helwan observatory near Cairo, Egypt.

795. "Double Attraction in Sky Last Night--Comet and Eclipse to Be Seen." **Wichita Daily Times** [Wichita Falls, Texas], 24 May 1910, 1.

Column 2. Halley's Comet, which has held the center of the celestial stage for several days, had a rival last night from a lunar eclipse. The result was that the comet lost out almost entirely as almost everyone looked at the moon.

796. "Halley's Comet." **London Times**, 24 May 1910, 13.

Column e. Reports observations made at Hampstead; Mortimer, Berks; and Balleyrane Observatory, county Wexford.

797. "Luminous Horn Jets from the Face of the Comet." **Minneapolis Morning Tribune** [Minneapolis, Minnesota], 24 May 1910, 1.

The phenomenon was observed by Professor Leavenworth at the University of Minnesota. A bright sector extends some distance ahead of the comet's main body. The theory is that the sun is shining on only one side of the comet thereby causing volatilization.

798. "No Eclipse and No Comet." **New York Times**, 24 May 1910, 6.

Column 2. Heavy fog obscures the comet from the view of local New York City astronomers.

799. "Philadelphia Sees Eclipse." **New York Times**, 24 May 1910, 6.

Column 2. Heavy clouds, however, blot out any view of Halley's Comet."

800. "See the Comet? Also the Eclipse." **Topeka Daily Capital** [Topeka, Kansas], 24 May 1910, 1.

Column 4. Describes the observation of the lunar eclipse between 9:40 P.M. and 1:22 A.M. at Washburn College in Topeka, Kansas. The comet was clearly visible during the eclipse.

801. "Taft Sees the Comet." **New York Times**, 24 May 1910, 6.

Column 2. President Taft views comet at the U.S. Naval Observatory.

802. "The Comet and Its Tail." **New York Times**, 24 May

1910, 8.

Column 3. Editorial. A satirical explanation of why the tail preceded comet in the eastern sky but followed it in the western sky.

803. "Thousand See Great Comet from Seattle." **Seattle Times** [Seattle, Washington], 24 May 1910, 2.

Columns 4-6. The heavenly visitor, gleaming like the headlight of a colossal locomotive, works its fiery way through the western sky. The tail resembles a gigantic yellow fan almost closed, extending from a head that twinkles and scintillates.

804. "Too Many Comet Reports." **New York Times,** 24 May 1910, 6.

Column 2. Navigators' zeal in wiring observations has unexpectedly overburdened the Hydrographer's Office. Reports in response to the agency's earlier request (see entry 440) for observation all come collect and are straining the budget allocation for telegraph and cable expenses.

805. "Two Astronomical Phenomena at Once." **Rapid City Daily Journal** [Rapid City, South Dakota], 24 May 1910, 1.

Column 1. Students at the South Dakota School of Mines have embraced an opportunity of a lifetime viewing both the comet and the eclipse occurring at the same time. As the eclipse became total the comet in the west showed more plainly, but was not brilliant although its tail extended far into heavens.

806. "Two Ring Circus in Sky at Night." **Register and Leader** [Des Moines, Iowa], 24 May 1910, 1.

Column 3. With an enthusiastic audience of millions, the two ring circus scheduled by the planets last night was put on without a hitch. Luna played hide and seek with old Sol while Halley's Comet, the only living comet in near captivity, competed for attention.

807. R., M.H. "Miss Proctor's Success." **New York Times,** 24 May 1910, 8.

Column 5. Letter lauds Mary Proctor's reports on the comet, referring to comets as the most mysterious bodies in our solar system. Perhaps they are created to sweep the realms of space.

808. "Comet Dimly Seen." **New York Times,** 25 May 1910, 2.

Column 3. A large meteor trailing bluish light flared across the sky near the comet--northward.

809. "Light Stream from Comet." **New York Times,** 25 May 1910, 2.

Column 3. The Harvard Observatory notes that the comet shows a jet toward the sun--its tail away from sun. The total light from the comet was set at 2.5 magnitude.

810. "The Comet." **Houston Post** [Houston, Texas], 25 May 1910, 6.

Poem. "It posed and made/A lot of fuss,/And then it took/A shot at us./And then it went/A-scooting by--/A yellow streak/Against the sky!/It must have moved/Roosevelt to mirth,/A thing which couldn't/Hit the earth."

811. "Turkish Fears of Halley's Comet." **London Times,** 25 May 1910, 7.

Column c. In Constantinople municipal authorities used the comet as a diversion and rounded up the cities stray dogs while a superstitious populace sought refuge from the comet. The people were on edge and cheered when morning arrived without disaster.

812. Raven-Hill, L. "The Great Amateur." **Punch,** (25 May 1910), 381.

Cartoon. Caption reads: " Aviator. "Marvelous Flier! And does it for love!" An aviator on ground leans against his airplane and looks admiringly skyward at Halley's Comet soaring by.

813. "Diminishing Comet Hidden." **New York Times,** 26 May 1910, 2.

The comet is hidden from New York City by a mist filled sky. By the end of the week it will be almost invisible to the naked eye. Reports other observations from Yerkes Observatory and the University of Arizona.

814. "Feared World's End in Constantinople." **New York Times,** 26 May 1910, 6.

Column 1. Constantinople inhabitants reportedly expected death when the comet approached. All night vigils were held. The municipal government took the opportunity to round up city's stray dogs without interference from people on streets.

815. "Halley's Comet." **London Times,** 26 May 1910, 11.

Column b. During a discussion of the comet at a meeting of the British Astronomical Association there was divergence of opinion as to whether the tail had been seen, either as a single or double object, or at all.

816. "Observations on Halley's Comet." **Nature**, 83 (26 May 1910), 384-86.

Reports on observations at Madrid Observatory and Malta. 2 photographs and spectrum are included. Also numerous shorter reports from around the world--Helwan, Transvaal, Kodaikanal, and Meudon.

817. "People See Comet and Eclipse." **Ward County Independent** [Minot, North Dakota], 26 May 1910, 3.

"It isn't often that one will see a comet and an eclipse of the moon in the same night." The eclipse started shortly past 9 o'clock and was almost complete by 11 o'clock--the comet was visible until nearly midnight. The sight was witnessed by nearly everyone in Minot.

818. "Saw Halley's Comet 75 Years Ago." **People's Banner** [Davis City, Nebraska], 26 May 1910, 7.

Column 1. An 85-year old resident reminisces about seeing the comet in 1835 when a boy in Minnesota. The comet appeared to be a bright star with a tail about a foot long. Seeing the comet sparked his lifelong interest in astronomy. He views comet as of no use except as a show.

819. Chree, C. "Halley's Comet and Magnetic and Electrical Phenomena." **Nature**, 83 (26 May 1910), 367-68.

Letter. Suggests that in his observations of the comet there was nothing in the electrical phenomena that was not adequately accounted for in the present meteorological conditions.

820. Eddington, A.S. "Halley's Observations on Halley's Comet, 1682." **Nature**, 83 (26 May 1910), 372-73.

Transcribes entries from a notebook found in the record room of the Greenwich Observatory. The surviving pages record observations made during late August and early September 1682.

821. Porter, Elizabeth Crane. "Curiosity and the Comet." **Nation**, 90 (26 May 1910), 535.

Letter. Relates poem about comet that she found in a 1759 newspaper clipping. The poem is reprinted in its entirety.

822. Turner, H.H. "Halley and His Comet." **Nature**, 83 (26 May 1910), 387-88.

Abstracted from the Alfred Lecture delivered before the Royal Society of Arts on 4 May 1910. Presents the history of Halley's computational analysis for predicting the periodicity of the comet.

823. "Good View at Harvard." **New York Times**, 27 May 1910, 5.

Column 4. The projection recently observed protruding from the comet's head has disappeared. The comet's magnitude is 6.43 as compared with the 6.57 of two nights ago. The head of the comet is sharply defined and star-like.

824. "Oh, What Fools We Mortals Be!" **Washington Standard** [Olympia, Washington], 27 May 1910, 2.

Column 1. Reports that in Aline, Oklahoma a girl was being held on an altar by religious fanatics. An occasional comet serves to remind man of his littleness in his efforts to grasp the works of the Almighty--except they be controlled by fixed laws.

825. Proctor, Mary. "Comet's Vagaries May Mark Its End." **New York Times**, 27 May 1910, 5.

Columns 2-4. There is evidence that Halley's Comet is breaking up, similar to others that have disappeared. In 1986 it may be twin comets that appear. But in any case, it is only a matter of time until the comet goes to pieces.

826. "Comet Gazers Throng Riverside Drive." **New York Times**, 28 May 1910, 2.

Column 4. An excellent view was available last evening and the avenue was thronged with curious spectators. Many people spent the entire evening in the open air gazing skyward despite the chilly weather.

827. "The Comet As a Puzzle." **Outlook**, 95 (28 May 1910), 133-34.

Mentions that observation at Manila and South Africa indicate the absence of solid matter even at the nucleus. Emphasizes how little is known about comets.

828. "The Comet." **Literary Digest**, 40 (28 May 1910), 1071.

Comments that in the aftermath of the comet's passing, predictions of some scientists were so very precise and so very wrong. In the main, however, scientific opinion lent little support to sensational expectations. "New York felt it was 'buncoed'."

829. "The Transit of Halley's Comet." **Scientific American,**

102 (28 May 1910), 438.

The doubtfulness of the earth's passing through the tail of the comet on 18 May proves once again what may happen to the best laid plans of mathematicians. The tail of Halley's Comet has conducted itself in a most whimsical fashion.

830. Proctor, Mary. "Bright View of the Comet." **New York Times**, 28 May 1910, 2.

Column 4. Clear skies reveal the comet more distinctly than at any time since 20 May. Only 2 or 3 degrees of tail are visible.

831. "Pope Not Impressed By Halley's Comet." **New York Times**, 29 May 1910, IV, 2.

Column 6. Pope Pius X says that for what one could see of the comet, it does not justify the commotion it has caused.

832. Proctor, Mary. "Comet's Tail Longer." **New York Times**, 29 May 1910, II, 13.

Column 6. The tail could be seen for 10 degrees from its nucleus last night. Observation conditions were poor because of mist.

833. "Clouds Hide the Comet." **New York Times**, 30 May 1910, 7.

Column 5. The sky was too hazy over New York last night to make viewing of the comet possible.

834. Forbes, George. "Halley's Comet." **London Times**, 31 May 1910, 10.

Column a. Letter. Dated 24 May. Having been at sea the last eleven days, the author has received no information from other quarters of observation about the comet. He provides his own observations in detail for 14 through 20 May 1910.

835. Proctor, Mary. "Comet Again Under Clouds." **New York Times**, 31 May 1910, 5.

A storm hid the comet from view last night. Moonlight will interfere with observation after 10 June, but until that time it will be possible, weather permitting, to obtain a view of the comet after sunset.

836. "Comet Notes." **Journal of the British Astronomical Association**, 20 (June 1910), 498-508.

Reports various observations of the comet at

Johannesburg, Trinidad, Jamaica, Melbourne, Accra, Calcutta, Poona, Providence (Rhode Island), Bloemfontein, Cape Town, Montpellier, and Torquay. The comet was not seen in England later than June 14.

837. "Notes." **Observatory**, 33 (June 1910), 255-56.

Presents an ephemeris for August to December 1910 computed by F.E. Seagrave. States that Halley's Comet has not presented a very remarkable spectacle in the British Isles, but reports from southern stations indicate that it has sustained its reputation there.

838. ["Report of Meeting"]. **Journal of the British Astronomical Association**, 20 (June 1910), 462-64.

John Evershed presented a report with slides on the observation of the comet at the Kodaikanal Observatory. Also exhibited were some spectroheliograph photographs which seemed to indicate an increase in the rate of angular rotation with height above photosphere.

839. ["Report of the Meeting"]. **Journal of the British Astronomical Association**, 20 (June 1910), 466-67.

A.C.D. Crommelin reads observation reports, and comments that it was impossible to deal with all the observations received from all over the world. They brought out forcibly the fact that Halley's Comet had done itself full justice.

840. Aitken, R.G. "Visual Observations of Halley's Comet, January-May 1910." **Publications of the Astronomical Society of the Pacific**, 22 (June 1910), 134-36.

During the period of the comet's greatest brightness the instrumental equipment of the observatory was principally devoted to securing a complete as possible set of spectrographic, photographic, and polariscopic observations. The nucleus and coma were visually observed.

841. Antoniadi, E.M. "The Alleged Sunward Direction of the Tail of Halley's Comet." **Journal of the British Astronomical Association**, 20 (June 1910), 492-93.

Letter. Evidence does not support the contention that the tail ever appeared directed sunwards, since both the morning and evening tails were seen ending on the side most removed from the sun.

842. Barnard, E.E. "Photographic Observations of Halley's Comet." **Popular Astronomy**, 18 (June 1910), 321-22.

Photographs indicate that the comet is in no way a freak comet. A photograph taken on 3 May 1910 at

Yerkes Observatory is included.

843. Bond, A. Russell. "Halley's Comet." **St. Nicholas,** 37 (June 1910), 678-81.

It is lucky that these are not the Middle Ages or we should be having the fright of our lives. Geared for children, the article explains the periodicity of the comet and why there is no need to fear it. Diagrams and photographs are included.

844. Curtis, Heber D. "Photographs of Halley's Comet Made At the Lick Observatory." **Publications of the Astronomical Society of the Pacific,** 22 (June 1910), 117-30.

A major observation article. Based on photographs of comet taken on 95 nights (12 September 1909 through 7 July 1910) at the Lick Observatory, the nucleus, coma, tail, and passage through tail are described in great detail. Plates are included.

845. Denning, W.F. "Halley's Comet." **Observatory,** 33 (June 1910), 253-54.

From Bristol, England the best view of the comet was on 23 May 1910 at 10:00 A.M. through a two-inch field glass--the tail was traced over 6-7 degrees. Thousands of people congregated on elevated places to view the comet.

846. Ledger, Edmund. "Halley's Comet in 1682." **Journal of the British Astronomical Association,** 20 (June 1910), 491-92.

Letter. Brings to attention the recorded observations of Dr. Robert Hooke in 1682. Cites three drawings of the comet on 20, 26 and 30 August 1682 and quotes from Hooke's observations.

847. Mitchell, S. Alfred. "Evening Sky Map for June." **Monthly Evening Sky Map,** 4 (June 1910), 3.

The cover has a photograph of the comet taken by E.E. Barnard with a 10-inch Bruce lens at Yerkes Observatory. In spite of cloudy weather Halley's Comet turned out to be just as remarkable as it was expected to be. On May 13 a naked eye view showed the tail to be 35 degrees in length.

848. Phillips, T.E.R. "Observations of Halley's Comet." **Journal of the British Astronomical Association,** 20 (June 1910), 480-82.

The best view was obtained on 26 May. Photographs of the comet on 22, 26, and 31 May 1910 are provided.

849. Stein, J. "A Newly Discovered Document on Halley's Comet." **Observatory**, 33 (June 1910), 234-38.

Discusses a new document discovered in Vatican records by Msgr. Bevilacqua. The document represents the first European observation of Halley's Comet for 1066. The document states that the comet was visible in Italy for over 50 days--longest duration of visibility recorded for the 1066 apparition.

850. Wilson, Ralph E. "The Story of Halley and His Comet." **Popular Astronomy**, 18 (June 1910), 357-66.

The story of Halley's association with Newton and Newton's influence on Halley's calculations. Also reviews historical events concurring with the comet's apparitions and observations.

851. "Observations of Halley's Comet." **Nature**, 83 (2 June 1910), 409-10.

A synopsis of observations made at Helwan, Marseilles, Khartoum, Malta, and Boulogne-sur-Seine.

852. "Overrated." **Life**, 42 (2 June 1910), 1009.

The "comet was more trouble than it came to. It was no more than a side show at best. It gave its performance at very inconvenient hours. Notice to posterity: Don't get excited by Halley's Comet. It is hardly up to its advertisement."

853. ["Circling Hawk ...], **Guymon Herald** [Guymon, Oklahoma], 2 June 1910, 1.

Column 2. Circling Hawk, a Creek Indian nearly a century old, saw the comet and muttered a dire prophecy: "When the star with a tail came before it was because my people had been wicked. ... streams dried up, cattle and horses died, game went away. ..."

854. Saunders, T. Bailey. "Halley's Comet." **London Times**, 2 June 1910, 5.

Column e. Letter. "Before the public interest in Halley's Comet fades" the author wishes to bring public attention to a description of it at its appearance in 1531 written on 13 August of that year by Melanchthon from Wittenberg (provides quotation).

855. "Halley's Comet." **Nature**, 83 (9 June 1910), 439.

A synopsis of various observations made at Koenigstuhl, Breslau, Sutton, Surrey, Thessaly, Paris, and Sofia.

856. Dines, W.H. "Meteorological Observations During the

Passage of the Earth Through the Tail of Halley's Comet."
Nature, 83 (9 June 1910), 427.

Letter. The author has examined records from ten
registering balloons sent up on 18, 19, and 20 May
1910. Nearly all traces show large fluctuation in
temperature, but such has occurred before. The passage
of the earth through the comet's tail had no affect on
the temperature of the upper air.

857. Bisson, P. "Observations of Halley's Comet Made on
Board S.S. Mobile." **Monthly Notices of the Royal
Astronomical Society**, 70 (10 June 1910), 617.

No split in the tail was observed on 7 May 1910.

858. Chatwood, A.B. "The Transit of Halley's Comet."
Monthly Notices of the Royal Astronomical Society, 70 (10
June 1910), 614.

Extract of a letter from Chatwood to the Astronomer
Royal (W.H.M. Christie), dated 19 May 1910,
Secunderabad, Deccan. "Transit no trace comet seen."

859. David, H.P. and H.H. McGill. "Observations of
Halley's Comet Made on Board S.S. Romania." **Monthly
Notices of the Royal Astronomical Society**, 70 (10 June
1910), 618.

Observed 18 May 1910: "Little or no tail to be seen,
probably owing to the moon being at the full and
shining brightly."

860. Evershed, J. "Halley's Comet and Its Spectrum As
Observed at Kodaikanal." **Monthly Notices of the Royal
Astronomical Society**, 70 (10 June 1910), 605-10.

The comet was first observed as a morning star on 18
April--this is earlier than had been anticipated.
Describes the content of the spectrum and the
methodology used in obtaining results. Three plates
from observations on 22 April, 1 and 13 May 1910 are
included.

861. Evershed, J. "Observations of the Tail of Halley's
Comet Before and After the Day of Transit." **Monthly Notices
of the Royal Astronomical Society**, 70 (10 June 1910),
610-11.

During the first two weeks of May the comet presented a
fine spectacle on clear mornings, and had attracted
considerable public attention and interest. The
extraordinary persistence of light in the eastern sky
might be explained by a strongly curved tail.

862. Evershed, J. "The Transit of Halley's Comet Across

the Sun." **Monthly Notices of the Royal Astronomical Society**, 70 (10 June 1910), 612-13.

A report based on the following program of observation: 1) scrutiny of sun's disk by direct observation; 2) a series of direct photographs; 3) a series of monochromatic photographs by spectroheliograph; (4) photographs with camera slit on line K; and 5) photographs with a spectrograph.

863. Hall, Maxwell. "Halley's Comet, 1910: Observations Made in Jamaica." **Monthly Notices of the Royal Astronomical Society**, 70 (10 June 1910), 615-17.

Presents observations made on 6 May through 22 May 1910 near Kingston, Jamaica.

864. Jenkins, J. "Observations of Halley's Comet Made on Board S.S. Majestic." **Monthly Notices of the Royal Astronomical Society**, 70 (10 June 1910), 619.

Observed on 28 May 1910. The visible part of the tail was 30-degrees in length. The luminosity of the tail faded gradually and evenly as its distance from the head increased. The brightness of head and tail varied at times like the Northern Lights.

865. Neill, G.A. "Sextant Observations of Halley's Comet Made on Board S.S. Urquhart." **Monthly Notices of the Royal Astronomical Society**, 70 (10 June 1910), 620-21.

Observed on 27 April and 1, 3, 4, 5, 6, 7, 8, 11, 12, 13 May 1910.

866. "Further Observations of Halley's Comet." **Nature**, 83 (16 June 1910), 470-71.

A synopsis of observations of the comet at Koenigstuhl, Warsaw, Berlin, Spain, Teneriffe, Fabra, and Malta.

867. "The Transit and Tail of Halley's Comet." **Nature**, 83 (23 June 1910), 501-02.

Synopsis of reported observations of the tail of Halley's Comet.

868. Finlay, W.H. and W.A. Douglas Rudge. "The Tail of Halley's Comet on May 18-19." **Nature**, 83 (23 June 1910), 487.

Letter. Suggests that the earth may have divided the tail. Discusses the effects of the tail's polarization, changing from a luminous patch to a stream of light.

869. Payn, Howard. The Tail of Halley's Comet on May

18-19." **Nature,** 83 (23 June 1910), 487.

Letter. Estimates the tail to have been half-way across the sky. Suggests that the earth split the tail in two since he saw light at both ends of comet's path as it passed.

870. Griffith, W. Branford. "Halley's Comet." **London Times,** 27 June 1910, 12.

Column a. Letter. Dated 5 June, Accra, Gold Coast. "You at home do not seem to have been very successful in seeing Halley's Comet. Out here it was a glorious object. ..." Griffith's first thought was "how like a flaming sword" before remembering Josephus' reference to a star like a flaming sword over Jerusalem at its fall.

871. "Halley's Comet." **Nature,** 83 (30 June 1910), 534.

Synopsis of reported observations of the comet's tail at Transvaal, Aquila, Bloemfontein, and Fabra.

872. "Chinese Bombard Comet with Fireworks from Hilltops." **Popular Mechanics,** 14 (July 1910), 33.

During May a great wave of fear swept over the Chinese empire because of the comet's approach. As the comet approached the earth the Chinese began firing fireworks, taking deliberate aim at the visitor from almost every mountaintop.

873. "Sporting (Advance Wireless from Aurora Borealis)." **Observatory,** 33 (July 1910), 306.

A satirical vignette reprinted from the **Ottawa Evening Journal,** 19 May 1910.

874. ["Notes"]. **Observatory,** 33 (July 1910), 297-98.

Snippets of letters from correspondents who had observed the comet at Rangoon, Straits of Bonifacio near Marseilles, the Himalayas, and at the Imperial University Observatory, Varsovie, Russia.

875. ["Notes"]. **Observatory,** 33 (July 1910), 305.

An extract from a newspaper article in Stuttgart, Germany, 30 May 1910, concerning the passage of Halley's Comet over that city.

876. ["Notes"]. **Observatory,** 33 (July 1910), 305.

Reprint of an article, "What Is a Comet," from the **Egyptian Morning News,** 20 May 1910--explains in detail the behavior of the comet's tail.

877. Frost, Edwin B. "Observing a Comet: A Sketch of the World-wide Study of Halley's Comet." **World Today**, 19 (July 1910), 785-90.

Reviews many of the preparations for the observance of the comet's 1910 apparition in various parts of world. The comet was nearest to the earth on 20 May at a distance of 14,300,000 miles. Three photographs are included.

878. Millard, Bailey. "In the Year of the Comet." **Technical World**, 13 (July 1910), 487-96.

Hypothesizes that Halley's Comet has exerted influence on the meteorological conditions of the earth during the first half of 1910--causing ".evastating inundations, terrible typhoons, frightful and unseasonable frosts and snows, and unprecedented heat."

879. "Halley's Comet." **Nature**, 84 (7 July 1910), 19.

A synopsis of observations at Munich, Koenigstuhl, Athens, and Malta.

880. Moir, James. "Halley's Comet." **Nature**, 84 (7 July 1910), 9.

Letter. An attempt to photograph the comet without any special apparatus--seen in Pisces at 5:30 A.M., 17 May 1910 with 15' exposure. The tail was 90-degrees long on the 17th, 115 degrees on the 18th and only 15-20 degrees on the 20th, and 35 degrees on the 23rd of May.

881. Stenquist, D. and E. Petri "Earth-Current Observations in Stockholm During the Transit of Halley's Comet on May 19." **Nature**, 84 (7 July 1910), 9.

Letter. The measured current strength proved to be considerably above the normal at this time of day, though by no means reaching to that of a magnetic storm. See also entry 901.

882. "Halley's Comet." **Nature**, 84 (14 July 1910), 52.

A synopsis of observations made at Johannesburg, Mt. Wilson, Kodaikanal, and Madrid.

883. "Halley's Comet." **Nature**, 84 (21 July 1910), 86.

A synopsis of observations made at Teneriffe, Munich, and Santiago de Chile.

884. "Halley's Comet." **Nature**, 84 (28 July 1910), 120.

A synopsis of observations made at Pic du Midi. A photograph of the comet taken on 31 May 1910 shows a secondary nucleus at a distance of 17" from the primary nucleus.

885. "A Tale of the Comet." **Popular Mechanics,** 14 (August 1910), 189.

Relates the story of a devious shrine keeper in Japan who used the predicted appearance of the comet to bilk pilgrims into presenting offerings. When nothing happened to the earth as the comet appeared villagers subjected the shrine keeper to a severe beating.

886. ["Notes"]. **Observatory,** 33 (August 1910), 342.

A translation of a letter addressed to the Egyptian publication **Al Aram,** 18 December 1909, asking questions about the photographs taken at Helwan on 24 August, for the purpose of comparing those photographs with those taken by the correspondent on 23 June.

887. ["Notes"]. **Observatory,** 33 (August 1910), 337.

A extract from a speech by the High Master of St. Paul's School on Speech Day (as reported in the **Daily Telegraph,** 28 July 1910): "Only one Pauline had in a sense been a disappointment in the past year--Edward [sic] Halley. Hardly anybody saw his comet, and those who did, did not think much of it."

888. ["Notes"]. **Observatory,** 33 (August 1910), 341-42.

A poem by Victor Hugo about Halley's Comet.

889. Abell, E.W. "Meteors from Halley's Comet on May 6." **Popular Astronomy,** 18 (August 1910), 422-24.

The earth was probably hit by fragments from the comet on 6 May, and may continue to be so struck every year on the same date until the main body of the comet visits 76 years hence.

890. "Halley's Comet." **American Journal of Science,** 180 (August 1910), 154-56.

Review of Chambers' brochure published in 1910 by Clarendon Press: Oxford (see entry 207). Summarizes the foci of observation.

891. Leon, Luis G. "Halley's Comet in Mexico." **Popular Astronomy,** 18 (August 1910), 428-29.

Observed at the National Observatory of Tacubaya, April, May, and June, 1910.

892. Metcalf, Joel H. "Observations of Halley's Comet." **Popular Astronomy,** 18 (August 1910), 433.

Observations of the comet's tail on 16, 17, and 19 May. At Taunton, Massachusetts the comet bore so slight an inclination to ours that we would only rarely be well situated to observe its structure on the plane.

893. Moorehouse, D.W. "Halley's Comet." **Popular Astronomy**, 18 (August 1910), 426-27.

Letter. The author's observations of 20, 24, 29 and 30 May and 5 June 1910 are reported. Three photographs of the comet.

894. Seagrave, F.E. "Halley's Comet, May 18-20." **Popular Astronomy**, 18 (August 1910), 431.

Seagrave saw no effects caused by the passage of the earth through the tail of the comet on either the night of 18 or 19 May. Observations of the comet since it has been visible in evening sky show that it is moving almost exactly in its computed orbit.

895. Sperra, William E. "Comet Halley." **Popular Astronomy**, 18 (August 1910), 427-28.

Letter. Reports observation of the comet on 4, 11, 18, 19, 21, 22, 26, 27, 28, May and 5, 15, 24 June 1910 at Cleveland, Ohio.

896. Swartz, Helen M. "Observations of Halley's Comet." **Popular Astronomy**, 18 (August 1908), 431-32.

Reports observation of the comet from Vassar College Observatory on 13, 16, 17 and 19 May. The focus of the report is the comet's tail.

897. Crawford, R.T. "Perturbations of Halley's Comet Due to Action of Views and the Earth." **Astronomical Journal**, 26 (29 August 1910), 134.

Reports perturbations computed by Encke's method of special perturbations on the basis of elements published in **Lick Observatory Bulletin**, No. 179. They are for heliocentric rectangular elliptical coordinates.

898. Leuscher, A.O. "Note on the Predictions Regarding the Transit of Halley's Comet, 1910 May 18." **Astronomical Journal**, 26 (29 August 1910), 135-36.

Describes methods on which the predictions are based. It was found that the earth would pass 1/20 of an astronomical unit below the axis of the tail. The diameter of the tail is less than one million miles at the point where the earth passed closest to the comet.

899. Farris, R.L. "Magnetic Observations at Cheltenham,

Maryland, May 15 to 20, 1910." **Terrestrial Magnetism**, 15 (September 1910), 163-65.

Special registration of three magnetic elements were made at the Coast and Geodetic Survey observatory at Cheltenham, Maryland in response to Dr. Nippoldt of the Potsdam Magnetic Observatory.

900. Stebbins, Joel. "The Brightness of Halley's Comet As Measured with a Selenium Photometer." **Astrophysical Journal**, 32 (September 1910), 179-82.

As determined with the selenium photometer, Halley's Comet became brighter than first magnitude, but at no time when observed did it reach magnitude 0. It was observed during May 1910.

901. Stenqvist, D. and E. Petri. "Observations of Earth-Currents in Stockholm on May 19, 1910, During Passage of Halley's Comet." **Terrestrial Magnetism**, 15 (September 1910), 159-61.

Presents a record of the magnetic observations made by the Meteorlogiska Centralanstalten, Stockholm. Includes graph and tabular data. See also entry 881.

902. Tittmann, O.H. and J.E. Burbank. "Principal Magnetic Storms Recorded at the Cheltenham Magnetic Observatory, April 1 to June 30, 1910." **Terrestrial Magnetism**, 15 (September 1910), 168.

Tabular presentation of twelve observations.

903. **London Times**, 8 September 1910, 7.

Column a. W.H. Dines reports observations of the upper atmosphere made on 18-20 May via balloons which recorded exceptions and sudden changes in temperature. He is unwilling to yet attribute this to the proximity of Halley's Comet.

904. "Halley's Comet." **Nature**, 84 (8 September 1910), 322-23.

A synopsis of observations made at Tokyo, Sonnwendstein, and Tashkent.

905. Hedrick, John T. "The Sky for the Night of May 18-19." **Astronomical Journal**, 26 (13 September 1910), 140.

Report of observations made at Georgetown College Observatory. On the night of 18-19 May the sky at an altitude of 20 degrees was a darkish red in the north, but lighter in the east over the city of Washington. The color was obvious but not striking.

906. Lowell, Percival. "Velocities of Particles in the Tail of Halley's Comet." **Astronomical Journal**, 26 (13 September 1910), 141-43.

Photographs taken at Lowell Observatory furnish experimental evidence of the mode of genesis of the tail. Particles in the tail are moving under the action of repulsive force directed from the sun. Measures distance in the tail from the nucleus. See also entries 926 and 948.

907. Perrine, C.D. "The Passing of the Earth and the Tail of Halley's Comet." **Astronomical Journal**, 26 (13 September 1910), 145-46.

Observed 17, 18, 19, 20, 21 May 1910 at Cordoba, Argentina. The width of the tail is widest near the horizon. Two branches of the tail are clearly extending from the horizon up to the Milky Way.

908. Whitney, Mary W. "Tail of Halley's Comet." **Astronomical Journal**, 26 (13 September 1910), 140.

Reports observations made at the Vassar College observatory. The comet's tail tapered away from the sun instead of growing broader.

909. Wilson, H.C. and R.E. Wilson. "Observations of Halley's Comet." **Astronomical Journal**, 26 (13 September 1910), 141.

Observations made at Goodsell Observatory, Carleton College, Northfield, Minnesota with a 16-inch refractor telescope. Included are two tables showing the locations of the comet and mean places of comparison stars.

910. "Further Observations of Halley's Comet." **Nature**, 84 (22 September 1910), 374.

A synopsis of observations made at Kodaikanal, Cordoba, Argentina, and Teneriffe.

911. "Velocities and Acceleration of the Ejecta from Halley's Comet." **Nature**, 84 (29 September 1910), 404.

Measuring photographs of the comet taken at Yerkes Observatory, Honolulu, and Beruit on 6 June 1910, Prof. Barnard has found that velocity of recession is a well marked feature of the comet's tail. Tabular results that are included show a strong acceleration in mass measured.

912. "At Odds on Halley's Comet." **Observatory**, 33 (October 1910), 417.

Astronomers meeting at Cambridge, England disagree on the comet's composition. Ellerman and Barnard of the U.S. regard the tail as composed of cyanogan, while A. Fowler of London maintains carbon monoxide (see entry 978) is pre-eminent in tail.

913. "[Note on] Halley's Comet." **Publications of the Astronomical Society of the Pacific,** 22 (October 1910), 195-97.

About a discussion of the comet at a meeting of the Royal Society of Edinburgh in reaction to a presentation given by Dr. John Aitken titled "Did the Tail of Halley's Comet Affect the Earth's Atmosphere?"

914. Ellerman, Ferdinand. "An Expedition to Photograph a Comet." **Publications of the Astronomical Society of the Pacific,** 22 (October 1910), 165-68.

The Astronomical and Astrophysical Society of America sent an expedition headed by Ellerman to Hawaii in order to better observe the comet. The expedition was funded with support from a grant received from the Rumford Fund of the National Academy of Sciences in Autumn 1909. Explains preparations for observation and weather conditions. Photographs included.

915. McGillivray, D. "The Chinese and the Comet." **Contemporary Review,** 98 (October 1910), 409-12.

In 1909-10 a comet poster for propaganda purposes was prepared by the Christian Literature Society in China. Tract Societies also prepared literature--over 115,000 copies sold. The result was relative calm among the populace and more respect for foreigners' knowledge.

916. Merfield, C.J. "Halley's Comet." **Journal of the British Astronomical Association,** 21 (October 1910), 46-47.

Letter. Merfield attests to the fact that the mean motion of a long-period comet cannot possibly be found with great accuracy from observation extending over a short period of time.

917. O'Halloran, Rose. "Observations of Halley's Comet." **Popular Astronomy,** 18 (October 1910), 453-56.

A chronological summary of observation during the comet's current apparition. The head of the long-heralded traveler was an object of thrilling interest.

918. Wilson, H.C. "Sketches on the Jets In the Head of Halley's Comet." **Popular Astronomy,** 18 (October 1910), 477-79.

The author has measured the position angles of the jets which apparently issued from the nucleus and has made hasty sketches of them as they appeared in the 16-inch telescope of Goodsell Observatory at Carleton College.

919. Todd, David. "Halley's Meteors." **Nature**, 84 (6 October 1910), 439.

Letter. From Johannesburg, South Africa. Relates the importance of obtaining observations in order to establish the connection between meteors and comets. Suggests that conditions will be best on 18 October 1910 for such observation because the earth then will be closest to orbit of the comet.

920. ["Report of Meeting"]. **Journal of the British Astronomical Association**, 21 (November 1910), 86-87.

A.C.D. Crommelin reports on various observations of the comet; also exhibits photographs taken by Winton Goatcher, of Cape Colony, with a small apparatus with a 3-inch aperture from an elevation of 2000 feet.

921. Innes, R.T.A. "Brightness of Halley's Comet." **Observatory**, 33 (November 1910), 444.

Letter. Dated Transvaal Observatory, Johannesburg, 26 September 1910. Clarifies the statement that Halley's Comet was brighter than Venus. The mass-impression of senses may say it was greater, but facts offered do not support such a statement.

922. Perrine, C.D. "Photographs of Halley's Comet Made at the Cordoba Observatory, Argentine Republic." **Monthly Notices of the Royal Astronomical Society**, 71 (11 November 1910), 102-03.

Extract from letter. It is clear that the bright streamer of the tail was not encountered by the earth, but passed to the north of us. There was, however, a wide, faint streamer to the south which, if it persisted, should have encountered the earth.

923. "Comet Notes." **Journal of the British Astronomical Association**, 21 (December 1910), 164-65.

Halley's Comet has been reobserved during November and December at Yerkes, Helwan, Nice, and Algiers--its magnitude is about 14. The comet will be approaching the earth until 9 February, therefore observation is expected for two or three more months.

924. Crommelin, A.C.D. ["Notes"]. **Observatory**, 33 (December 1910), 491-92.

Halley's Comet is now sufficiently emerged from the

sun's rays to give ground for the hope of a few more observations. Presents an ephemeris for December 1910 through March 1911.

925. Perrine, C.D. "Resume of Observations of Halley's Comet at Cordoba." **Publications of the Astronomical Society of the Pacific,** 22 (December 1910), 211-13.

States that, theoretically, Halley's Comet became bright enough to observe with 12-inch equatorial telescope on about 1 December 1909. It was first looked for on 30 November and found much brighter than expected. Describes observations through May 1910.

926. "The Motion of Molecules in the Tail of Halley's Comet." **Scientific American,** 103 (3 December 1910), 439.

A synopsis of Lowell's bulletin on the motion of molecules in the tail of Halley's Comet; includes table. See also entries 906 and 948.

927. "Recent Helwan Photographs of Halley's Comet." **Nature,** 85 (8 December 1910), 180.

Abstracted from **Astronomische Nachrichten,** no. 4457. Presents recent photographs taken with the Reynolds reflector at Helwan Observatory on 7 and 9 November 1910 which indicate a correction of +0.2 m, 0' to the ephemeris published in **AN,** No. 4450. The visual brightness is greater than previously thought.

928. Rambaut, A.A. "Photographic and Visual Observations of Halley's Comet (1909c), Daniel's Comet (1909e), and Comet 1910a, Made at the Radcliffe Observatory, Oxford." **Monthly Notices of the Royal Astronomical Society,** 71 (9 December 1910), 133-41.

Contains tabular and narrative results of observations (visual and photographic) of Halley's Comet after the date of perihelion passage.

929. Humphreys, W.J. "On Passing Through Tail of Halley's Comet." **Bulletin of Mount Weather Observatory,** 3 (17 December 1910), 239-44.

The scientific world expected only slight terrestrial disturbances from the comet, and does not feel certain that any effect at all was observed. Lists phenomena possibly due, in part, to Halley's Comet which were observed by U.S. Weather Bureau officials on 19 May. See also entry 976.

930. Raymond, W.E. "Transit of Halley's Comet, 1910 May 19. Observations Made at Sydney Observatory." **Monthly Notices of the Royal Astronomical Society,** 70 (Supplement 1910), 653.

Report of observation made on 19 May 1910.

1911

931. **American Yearbook 1910.** New York: Appleton, 1911, pp. 559-60.

   Summary of observations of the comet. Although no comment is made on the social reaction or disruption caused by the comet, the piece is interesting in that it sets the comet's passage in the context of the year's events.

932. Butler, Charles P. "The Spectrum of Halley's Comet." **Proceedings of the Royal Society of London,** 84 (1911), 523-26.

   Read 24 November 1910. On 23 and 26 May visual observations of the spectrum were secured. Brightness was of the second magnitude, much less than expected. Describes the spectrum of the coma and nucleus. The tail, however, was not observed due to bad weather.

933. Slipher, V.M. "The Spectrum of Halley's Comet in 1910 As Observed at Lowell Observatory." **Lowell Observatory Bulletin,** 2, No. 2 (1911), 3-16.

   Gives a lengthy list of split spectrograms and notes descriptive of the split spectrograms. There are also descriptions of the plates, five pages of which are included, illustrating various spectral images.

934. "Comet Notes." **Journal of the British Astronomical Association,** 21 (January 1911), 203.

   Reports observation by Prof. Barnard on 8 January 1911--the magnitude of the comet is 13 or 14. Barnard hopes to follow the comet with a 40-inch refractor telescope for the greater part of 1911. Also reports an observation by Cereski at Moscow while the comet was in transit.

935. "Halley's Comet." **Nature,** 85 (12 January 1911), 349-51.

   Provides a synopsis of observations of the comet at Transvaal, Buluwayo, Utrecht, Algiers, and Teneriffe. There is a lengthier discussion of the Transvaal observation since it is viewed as being better situated for clear and protracted viewing of the comet's passage.

936. Tebbutt, John. "Observations of Halley's Comet at Windsor, N.S. Wales." **Monthly Notices of the Royal**

Astronomical Society, 71 (13 January 1911), 224-26.

Observed 3 December 1909 through 1 July 1910. Gives a table of comparison stars. The comet's nucleus was faint in December and subject to small fluctuations in brightness. On the morning of 22 April the tail could be traced for 2 or 3 degrees from the head.

937. "Comet Notes." **Journal of the British Astronomical Association,** 21 (February 1911), 244.

Reports observation by Prof. Barnard, as well as an observation from Cape Town with a 12-inch refractor telescope on 5 February 1911. It is likely that the comet will be followed at least until its next conjunction with the sun which will occur about August.

938. "The Chinese and the Comet." **Missionary Review of the World,** 24 (February 1911), 142.

A striking example of the influence of the Christian Literature Society of China was afforded by its propaganda to enlighten the Chinese regarding Halley's Comet--277,000 posters were circulated. As a result of these efforts the usual disturbances among superstitious people were absent entirely.

939. Innes, R.T.A. "Passage of the Earth Through the Tail of Halley's Comet." **Journal of the British Astronomical Association,** 21 (February 1910), 239-40.

Letter. Maintains that there was no passage of the earth through the tail of Halley's Comet.

940. Ross, David. "Halley's Comet." **Journal of the British Astronomical Association,** 21 (February 1911), 240-41.

Letter. Halley's Comet was watched in Australia with great enthusiasm. Many expected a better display, but weather conditions were not very favorable and the moon was often in the way.

941. Warner, Irene E. Toye. "Halley's Comet: The Romance of Its Past." **Popular Astronomy,** 19 (February 1911), 103-06.

A poem precedes this essay. Many of us who expected to behold a most brilliant fiery object with a tail like a train of light amongst the stars have been disappointed. If regarded in a proper way, the comet replaces disappointment with admiration and awe.

942. "Comets of 1910." **Monthly Notices of the Royal Astronomical Society,** 71 (10 February 1911), 318-20.

The comet is still visible in large telescopes, but the last published observation is dated 7 December 1910.

943. B., R.S. "Orbit of Halley's Comet." **Monthly Notices of the Royal Astronomical Society**, 71 (10 February 1911), 320-23.

Discusses the methodology used by Cowell and Crommelin in their **Essay on the Return of Halley's Comet** (see entry 297). Gives a step by step explanation of how they mathematically calculated a prediction of the exact date of the comet's 1910 perihelion.

944. Beattie, E.H. "Halley's Comet." **Journal of the British Astronomical Association**, 21 (March 1911), 269-71.

Letter. Takes issue with the premise that the earth passed through the comet's tail. Also finds it difficult to accept the statement that the nucleus was at its brightest no more than second magnitude, since it was much brighter than Saturn.

945. Horner, G.R. "Observations of Halley's Comet." **Journal of the British Astronomical Association**, 21 (March 1911), 271-72.

Letter. Reports observations in New Zealand during April, May, and June 1910. Nothing was seen of a secondary tail.

946. Astronomer Royal [W.H.M. Christie] "Observations of Comets Made at the Royal Observatory, Greenwich, in 1909-1910." **Monthly Notices of the Royal Astronomical Society**, 71 (10 March 1911), 475-78.

Presents position tables for Halley Comet for 9 September 1909 through 2 June 1910.

947. Keeling, B.F.E. "Observation of Halley's Comet." **Observatory**, 34 (April 1911), 161.

Letter. Dated 19 March 1911. Keeling is superintendent of Khedival Observatory, Egypt. The comet is still under observation here and has recently shown a decided increase in brightness. Throughout, the comet has been brighter visually than it has been photographically.

948. Lowell, Percival. "Spectroscopic Proof of the Repulsion by the Sun of Gaseous Molecules in the Tail of Halley's Comet." **Proceedings of the American Philosophical Society**, 50 (May 1911), 254-60.

Read 21 April 1911. The return of Halley's Comet has been noteworthy chiefly for the possibility of employing upon it modern methods of research. Since its last apparition spectroscopy and celestial photography have been developed. See entries 906 and 926 for less developed discussion by Lowell on the same topic.

949. Slocum, Frederick. "Halley's Comet." **Popular Astronomy,** 19 (May 1911), 282-83.

Discusses how much longer the comet will be able to be followed by observation. As of 28 March 1911 it has been followed for 249 days before perihelion passage and for over 300 days since.

950. Perrine, C.D. "Observations of Halley's Comet." **Astronomical Journal,** 27 (12 May 1911), 14-16.

Observed at Cordoba, Argentina with a 12-inch refractor telescope. Tables are presented which give the position of comet and mean places of comparison stars.

951. Shaw, H. Knox. "Positions of Halley's Comet and of Comet 1910a from Photographs Taken at the Khedivial Observatory, Helwan." **Monthly Notices of the Royal Astronomical Society,** 71 (12 May 1911), 573-77.

Describes the method of photography and measurement employed, giving a tabular rendition of the positions. To obtain the comet's mean place it is necessary to apply correction for parallax, reapply correction for the earth's aberration, and to antedate for light-time.

952. "Halley's Comet Followed." **New York Times,** 29 May 1911, 1.

Column 6. The comet is being followed into distant space with the telescope at the Lick Observatory. It is now a little further than Jupiter is from the sun--about 300 million miles. Last week it was photographed in exactly its predicted location.

953. "Ephemeris of Halley's Comet." **Popular Astronomy,** 19 (June 1911), 368-69.

Presents an ephemeris for June 1911.

954. Aitken, R.G. "Observing Halley's Comet From the Summit of Mt. Whitney." **Proceedings of the Astronomical Society of the Pacific,** 23 (October 1910), 240.

Communicates that G.F. Marsh climbed Mt. Whitney to observe Halley's Comet and the total eclipse of the moon on 23 May 1910. This observation is supposedly the highest point on earth from which the comet was observed.

955. Frost, Edwin B. "The Radial Velocity of Halley's

Comet As Derived from a Spectrogram." **Popular Astronomy,** 19 (November 1911), 558-59.

A spectrogram taken on 24 May 1910 indicated a radial velocity of +55 kilometers per second. This differs by less than a kilometer from the radial velocity inferred from the ephemeris of the comet as it was then receding from the earth.

956. Schwarzschild, K. and E. Kron. "Brightness in the Tail of Halley's Comet." **Astrophysical Journal,** 34 (December 1911), 342-52.

Measuring the suface intensity of the tail 1/2 degree behind the head it is deduced that 1/2 gram of matter flows per second from the head to the tail. The brightness of the tail diminishes continuously from the head of the comet outward, aside from irregularities. This phenomenon is caused by a falling off in the tail's density.

1912

957. Crommelin, A.C.D. and D. Smart. **Report of the Section for the Observation of Comets: Halley's Comet.** London: Eyre and Spottiswoode, 1912. 40 pp.

Includes data on the history of the comet, the prediction of its 1910 return, its brightness, length of tail in degrees, appearance of nucleus and coma, and the appearance and width of its tail. Reprinted in **Memoirs of the British Astronomical Association,** v. 19, part 1 (see entry 967).

958. "Comet Notes." **Journal of the British Astronomical Association,** 22 (January 1912), 204-05.

Presents a table of observations of brightness of the comet made at Helwan, the comet's revival reaching a magnitude of 1 about 25 February 1911. Goes on to discuss work on the connection between comets and meteors.

959. Curtis, H.D. "On the Distribution of Brightness in the Tail of Halley's Comet." **Proceedings of the Astronomical Society of the Pacific,** 24 (April 1912), 138-40.

Presents a synopsis of article by Schwarzschild and Kron that appeared in **Astrophysical Journal** (see entry 956). The data in the article was secured by the expedition from Potsdam Observatory to Teneriffe in the Spring of 1910.

960. ["Notes"]. **Observatory**, 35 (July 1912), 286.

A translation of an article by Dr. Hassan Fahmy Gamal
el Din which was published in **Al Aram**, 19 January 1910.
The comet appeared on the afternoon of 13 Jan 1910 at
5:45, its color is purple-red, somewhat bluish, and it
is rectangular in shape. The electricity of the
comet's center is positive and that of tail negative.

961. Olivier, Charles P. "The n Aquarid Meteor."
Astronomical Journal, 27 (24 September 1912), 129-30.

The major axis of aquarids is nearly the same as that
of Halley's Comet. That the elements do not agree more
closely is due principally to more of the meteors being
nearer than 5 million miles to the actual orbit of the
comet and some as far as 13 million miles away.

962. Turner, Herbert H. "Halley's Comet." **Proceedings of
the Royal Institution of Great Britain**, 19 (November 1912),
753-64.

A lecture presented at the weekly evening meeting of 18
February 1910. Discusses Halley's research and
predictions and Crommelin's orbit calculations and
searches for records of early sightings. Previously
reprinted in **Chemical News**, 101 (23 and 29 April 1910),
187 ff.

963. Barnard, E.E. "Micrometer Positions of Halley's
Comet, Made With the 40-inch Telescope of the Yerkes
Observatory." **Astronomical Journal**, 27 (16 December 1912),
147-52.

Presents tables giving the positions of the comet, the
mean places of comparison stars, measured positions of
comparison stars, measures of intermediate and other
stars, photographic positions of the comet, plus notes
on the comet at times of measurement (23 May 1910-29
May 1911).

1913

964. Curtis, H.D. "The Mass of the Nucleus of Halley's
Comet." **Proceedings of the Astronomical Society of the
Pacific**, 25 (June 1913), 175-77.

A synopsis of S. Orloff's paper which was published in
the **Bulletin of the St. Petersburg Academy**, No. 5,
1913. Just as there is a phase effect in light from
the moon, so also in the nucleus of a comet, which
consists either of a single mass or a swarm of small
particles.

183

965. Seagrave, F.E. "Ephemeris of Halley's Comet at Opposition in 1914." **Popular Astronomy,** 21 (October 1913), 513.

Surmises that at its next opposition, at about 9 February 1914, there will not be the slightest chance of seeing it. Provides an ephemeris for February 1914.

966. Toynbee, Paget. "Dante and Halley's Comet." **London Times,** 31 December 1913, 66.

Column b. Letter. Suggests that Dante, in his **Convivio,** recorded the 1301 appearance of Halley's Comet (quotes the passage to which he refers).

1914

967. Smart, D. and A.C.D. Crommelin. "First Report of the Comet Section: Halley's Comet." **Memoirs of the British Astronomical Association,** 19 (1914). 1-36.

Summary of overall attention devoted to the comet: History, predictions, return, observers, brightness, tail, nucleus and coma, spectra, transit of May 18, distance from the sun and earth, orbit diagrams, and photographs. Excellent summary of 1910 apparition. See also entry 957.

968. Blencoe, David A. "Variation in the Period of Halley's Comet." **Popular Astronomy,** 22 (March 1914), 191-93.

Letter. Two revolutions of the comet equal thirteen of Jupiter's. Calculates revolutions of the comet in relation to Jupiter, Saturn, Uranus, and Neptune as a potential explanation for the apparent cycle of ten of the comet's period intervals.

969. Evershed, Mary A. "Dante and Halley's Comet." **Journal of the British Astronomical Association,** 24 (April 1914), 362-63.

Letter. In regard to Paget Toynbee's letter in the **London Times,** (see entry 966), Evershed suggests that it is clear that Dante was referring to a meteor or shower of meteors, not a comet.

970. Barnard, Edward E. "Visual Observations of Halley's Comet In 1910." **Astrophysical Journal,** 39 (June 1914), 373-404.

Discusses Halley's Comet in lay terms, giving attention to the comet viewed with a large telescope, the velocity of particles in the tail, naked eye and telescopic observations, and the possible encounter of the earth with the comet's tail.

1915

971. Barnard, E.E. "Preliminary Report on the Photographs of Halley's Comet Taken at Honolulu, H.I., by Ferdinand Ellerman in 1910." **Publications of the Astronomical and Astrophysical Society of America**, 2 (1915), 66-67.

> Presented at the Society's 12th annual meeting, August 1911. Ellerman secured an excellent series of photographs and made very important, though negative, observations at the time of the comet's transit across the sun on 18 May 1910.

972. Comstock, George C. et al. "American Astronomical Society: Report of Comet Committee 1909-1913." **Publications of the Astronomical and 177-227 1915 Astrophysical Society of America**, 2 (1915), 177-227.

> An "Index-Catalogue of Photographs of Halley's Comet" comprises pp. 183-218 of the report; the photographs of Halley's Comet were taken by Ferdinand Ellerman at Diamond Head, Honolulu (28 plates). Notes on the Ellerman photographs are supplied by E.E. Barnard.

973. Curtis, H.D. "Photographs of Halley's Comet at the Lick Observatory." **Publications of the Astronomical and Astrophysical Society of America**, 2 (1915), 15.

> Presented at the 11th annual meeting, August 1910. Describes the method by which 370 negatives were taken of the comet between 12 September 1909 and 7 July 1910.

974. Ellerman, Ferdinand. "The Society's Expedition to Hawaii for Photographing Halley's Comet." **Publications of the Astronomical and Astrophysical Society of America**, 2 (1915), 5-16.

> Presented at 11th annual meeting, August 1910. Summarizes the expedition's intent and success as well as the equipment used for observation on the Society's observation expedition to Hawaii.

975. Frost, Edwin B. "Radial Velocity of Halley's Comet as Derived from a Spectrogram." **Publications of the Astronomical and Astrophysical Society of America**, 2 (1915), 69.

> Presented at the Society's 12th annual meeting, August 1911. The radial velocity of the comet on 24 May 1910 as determined from displacements of the Fraunhofer lines, was +55 km. per second.

976. Humphreys, W.J. "On Passing Through the Tail of Halley's Comet." **Publications of the Astronomical and Astrophysical Society of America**, 2 (1915), 35-38.

Presented at the Society's 11th annual meeting, August 1910. Lists halos, coronas, and other phenomena possibly due in part to Halley's Comet, and locations where they were observed on 19 May 1910. See also entry 929.

977. Barnard, E.E. "On the Acceleration of the Receding Masses in the Tail of Halley's Comet, June 6, 1910." **Publications of the Astronomical and Astrophysical Society of America**, 2 (1915), 17.

Presented at the Society's 11th annual meeting, August 1910. The opportunities here for determining the motion of the tail-forming particles of Halley's Comet have been few. Definite masses suitable for the purpose were seldom shown to the plates. It was too cloudy.

978. Fowler, A. "Photographs and Spectrum of Halley's Comet." **Publications of the Astronomical and Astrophysical Society of America**, 2 (1915), 18.

Presented at the Society's 11th annual meeting, August 1910. The photographs exhibited were taken by Evershed at Kodaikanal, India. Author gave an account of the experimental work that allowed him to identify bands of the tail with the spectrum of carbon monoxide.

1916

979. Very, Frank W. "What Is a Comet's Tail?" **Scientific American**, 114 (5 February 1916), 156 ff.

Presents some new results from the recent appearance of Halley's Comet. Discusses the visual effects of the tail and the possible scientific explanations for them. Includes two photographs and four diagrams.

980. "The Tail of Halley's Comet." **Observatory**, 39 (March 1916), 142.

Did the earth actually pass through the tail of Halley's Comet on 21 May 1910? M.E. Esclangon has calculated that actually the earth passed only 0.002 astronomical units outside the boundary of the tail at midnight on the 21st of May.

1917

981. Barnard, Edward E. "Note on the Southern Tail of Halley's Comet Seen May 17 and 18, 1910." **Astrophysical Journal**, 46 (July 1917), 83-84.

The second tail of the comet was in no way due to zodiacal light, but belongs entirely to the comet.

1919

982. Berlingett, Alice. "To Halley's Comet." **Popular Astronomy**, 27 (May 1919), 338.

Poem.

983. Hall, Maxwell. "Halley's Comet 1910." **Popular Astronomy**, 27 (May 1919), 281-87.

Observation from Kempshot Observatory, Jamaica. Presents a diagram of the orbits of the comet and earth, a table for plotting positions of earth and comet, and a drawing of the comet on 20 May 1910 at 4:00 and 7:30 A.M.

1921

984. "A Comet That Comes Back." **Mentor,** 9 (October 1910), 29.

"Of all the comets that blaze in the sky, there is only one that pays the earth regular visits." Halley's Comet is also said to have the most interesting history of any of these vagrants.

1926

985. Proctor, Mary. Return of Halley's Comet in 1910." In **The Romance of Comets.** New York: Harper, 1926. pp. 94-132.

Includes details of the author's own observations made from the summit of the Times building in New York City during May 1910. Miss Proctor was the prime correspondent covering the comet for the **New York Times.**

1927

986. "Gas of Halley's Comet Blamed for World War." **San Francisco Chronicle**, 7 May 1927, 1.

Column 7. Dr. F. Homer Curtiss, founder of the "Order of Christian Mystics," in addressing the Universal Women's Alliance, blamed the gaseous trails left in the earth's atmosphere by Halley's Comet for World War I--it made humanity nervous. See also entry 987.

987. "So This Is What Ails Our Suffering Globe!" **San Francisco Chronicle**, 9 May 1927, 18.

Column 1. A mystic prophet, F. Homer Curtiss, says that the world was gassed by Halley's Comet in 1910 and made mankind crazy. It is responsible for boll weevils, bolshevik coal strikes, prohibition agents, cover charges, taxes, Japan earthquakes, etc. See also entry 986.

988. Bobrovnikoff, N.T. "On the Spectrum of Halley's Comet." **Astrophysical Journal**, 66 (October 1927), 145-69.

Discusses sudden changes in the structure of the comet's head on 24 May 1910. Includes plates and tables showing the size, density, and intensity of spectral images of the comet.

1929

989. Bowles, Paul Frederic. "Halley's Comet." **This Quarter**, 1 (Spring 1929), 244.

Poem.

1930

990. "Halley's Comet." **Science**, 72 (29 August 1930), 10 supplement.

N.T. Bobrovnikoff has concluded that in 1910 the comet had two distinct tails which appeared as one when viewed from earth. One tail was a straight narrow streamer of carbon monoxide gas, while other was meteoric dust and more diffuse and curved.

1931

991. Bobrovnikoff, Nicholas T. "Halley's Comet in Its Apparition of 1909-1911." **Publications of the Lick Observatory,** 17, Part 2 (1931), 1-482.

Reviews the material and methods of measurement and plan of investigation. Presents a chronological description of the various cometary phenomena. Discusses the spectrum of Halley's Comet, the general structure of the comet, and comments that many of the thousands of papers about the comet published are worthless because of their lack of precision. Describes some of the vast amount of data accumulated by H.D. Curtis. Discusses the conception and content of his own study of the comet. This very important study was also published under the same title by the University of California Press (1931).

1932

992. Davidson, M. "Halley's Comet in Apparition of 1909-11." **Observatory,** 55 (October 1932), 282-89.

Reviews Bobrovnikoff's study (see entry 991). Concludes with Bobrovnikoff's thought that "what we actually observe may be but a link in the processes of interaction of cometary matter and solar energy."

1933

993. Crommelin, A.C.D. "Halley's Comet in 1910-11." **Nature,** 131 (25 February 1933), 282.

A review of N.T. Bobrovnikoff's "Halley's Comet in Its Apparition of 1909-1911" (see entry 991). Provides comment on the highlights of this important synthesis of observation and research.

1934

994. Tsu, Wen Shion. "Observations of Halley's Comet in Chinese History." **Popular Astronomy,** 42 (April 1934), 191-201.

An apparition by apparition (from 240 B.C. to 1910) discussion of Chinese records noting observations of the comet. Chinese calculations predict a February 1986 return.

1937

995. Proctor, Mary and A.C.D. Crommelin. **Comets: Their Origin, and Place in the Science of Astronomy.** London: Technical Press, 1937.

Two chapters, "The Story of Halley's Comet" (pp. 43-60), and "Return of Halley's Comet in 1910" (pp. 61-75). Both articles are synoptic in nature and provide a good overview of both astronomical and popular history, as well as reflecting the importance of Halley's Comet.

1939

996. Salmon, W.H. "Comets: History and Science." **Chambers's Journal,** 8 (June 1939), 425-28.

An overview article of comets throughout history. It is particularly interesting for the disproportionately large amount of space devoted to the historical and astronomical lore of Halley's Comet.

1942

997. Plummer, H.C. "Halley's Comet and Its Importance." **Nature,** 150 (29 August 1942), 249-57.

Reports on the Halley lecture delivered at Oxford, 25 May 1942. Provides insights into Newton's influence on Halley's work with the comet of 1682. Discusses what Halley's prediction and its fulfillment signified for the development of science.

1946

998. Wheeler, Robert L. "Halley's Threat." **Forum,** 105 (April 1946), 710-11.

Current speculation about atomic energy destroying world reminds the author of the 1910 Halley's Comet apparition when the world was being destroyed once a week in the Sunday supplement. Parodies 1910 hoopla. Reprinted from the **Providence Evening Bulletin.**

1955

999. Stephenson, Bill. "The Panic Over Halley's Comet." **Maclean's**, 68 (14 May 1955), 30 ff.

About the 1910 apparition of the comet when "sulphur rained down on Newfoundland, a mirage appeared over Niagara and scientists who should have known better scared the wits out of people all over the world." Includes photos of newspaper headlines and advertisements.

1000. Schove, D. Justin. "Halley's Comet, I. 1930 B.C. to A.D. 1986." **Journal of the British Astronomical Association**, 65 (July 1955), 285-89.

A review of the history of Halley's Comet astronomy. Mentions that Twain had commented that as he had been born while the comet was visible he would be greatly disappointed if he did not also go out with it.

1001. Schove, D. Justin. "The Comet of David and Halley's Comet." **Journal of the British Astronomical Association**, 65 (July 1955), 289-90.

The comet that King David saw, leading him to build an altar of atonement (the temple of Jerusalem), was Halley's Comet. Posits that the date that construction on the temple began was 989 B.C. Halley's Comet appeared in 1005 B.C.--since the temple was 16 years under construction it is chronologically possible.

1956

1002. Rexroth, Kenneth. "Halley's Comet." in **In Defense of the Earth**. New York: New Directions, 1956. p. 19.

Poem. The poem is one of a series of poems titled, "The Lights in the Sky are Stars."

1003. Alter, Dinsmore. "Comets and People." **Griffith Observer**, 20 (July 1956), 74-82.

About popular reactions the the 1910 return of Halley's Comet as reflected in contemporary newspaper accounts. The cover reproduces photographs of the comet taken by Ellerman at Honolulu in 1910.

1959

1004. Scott, Samuel. "The Reappearance [in 1759] of the

Celebrated Halley's Comet Depicted by a Painter of the Time, Samuel Scott." **Illustrated London News,** 235 (31 October 1959), II.

A color reproduction with caption, facing on p. 541, of one of the few paintings of the comet. A slightly smaller reproduction, also in color, appears in Sir Fred Hoyle's **Astronomy** (Doubleday, 1962) p. 139.

1960

1005. **New York Times,** 8 May 1960, 62.

Column 5. The role of Halley's Comet in Mark Twain's life is noted.

1006. Klein, Jerry. "When Halley's Comet Bemused the World." **New York Times Magazine,** 8 May 1960, 45 ff.

This article was written to mark the 50th anniversary of the 1910 apparition. Most of the facts given have been gleaned from the pages of the **New York Times's** comet coverage. The article mentions two 1910 cocktails inspired by the comet--"Cyanogen Cocktail" and the "Syzygy Fizz."

1962

1007. Yoke, Ho Peng. "Ancient and Medieval Observations of Comets and Novae in Chinese Sources." **Vistas in Astronomy.** ed. Arthur Beer. New York: Pergamon, 1962. pp. 127-225.

A year by year rendition of Chinese astronomical records. The years of Halley's Comet apparitions can be checked for appropriate observations of the comet. Yoke corrects previous work done on Chinese astronomical records (see entries 138 and 994).

1008. "New Comet May Become As Bright As Halley's." **Science News Letter,** 81 (23 March 1962), 185.

Comet Seki-Line may rival Halley's in brilliance. Halley's Comet is mentioned as one of only two first magnitude comets to have appeared this century.

1964

1009. Benton, Ray. "Exploring the Comets." **Spaceflight,** 6

(July 1964), 110-19.

In addition to a brief discussion of the history of comets, Benton discusses the general kinds of comet research being done and the need for the continuance of cometary research. Halley's Comet is mentioned frequently as a sort of archetypal comet. States that Ernst J. Oepik's calculation that the chance of the earth colliding with a comet 10 miles in diameter is 3 in 4.5 billion years--"hardly anything for us to worry about; yet, should the Earth encounter a comet nucleus of this size, the tremendous heat generated by atmospheric friction, shock wave and impact might possibly destroy all life in an area the size of North America."

1966

1010. Ivanov, K.G. and A.D. Shevnin. "Geomagnetic Phenomena Observed During the Transit of the Earth Through the Tail of Halley's 1910 II Comet." **Geomagnetism and Aeronomy,** 6, No. 5 (1966), 634-37.

The data of seven Northern Hemisphere observatories are used to analyze geomagnetic phenomena observed 18-19 May, 1910 during the transit of the earth through the tail of Halley's Comet. It is shown that at this time there was a small global geomagnetic field disturbance which was approximately symmetric with respect to the instant of passage of the comet over the solar disc. Some ideas are offered in favor of the existence of a geomagnetic effect from Halley's Comet.

1011. Mumford, George S. "Periodic Comet Halley." **Sky and Telescope,** 32 (August 1966), 71.

Notes that the comet is due to return in two decades and already astronomers are beginning to prepare for it. Expresses the need for an exact determination of the orbital elements back to 1835 and then forward from the 1986 apparition.

1967

1012. Richardson, Robert S. "Halley's Comet." in **Getting Acquainted With Comets.** New York: McGraw-Hill, 1967. pp. 196-209.

A good overview of the astronomical history of the comet. Provides illustrations of the orbit and earth position during the 1910 apparition.

1013. Sekanina, Z. and V. Vanysek. "Irregularities in the Motion of Comet Halley in 1910 and Its Physical Behavior." **Icarus,** 7, No. 2 (1967), 168-72.

Anomalous positional residuals from the orbit of the comet, ascertained by Zadunaiski can be explained by the existence of a spurious nucleus, deviating from the center of gravity of the comet (from its solid nucleus). It was moving behind the solid nucleus with a relative velocity of the order of a few meters per second. It can be understood as a slightly expanding cloud of micrometeoric grains of irregular shape.

1014. "Lockheed Study Shows Probe to Halley's Comet Is Feasible." **Aerospace Technology,** 21 (28 August 1967), 56.

The mission, made difficult because the comet is in a retrograde orbit, could be accomplished by a planetary swingby technique. The Lockheed plan envisions a rendezvous with the incoming comet from about 300 to 150 days before it reaches perihelion.

1015. Richardson, Robert S. "Return of Halley's Comet: 1985." **Science Digest,** 62 (November 1967), 70-75.

The return visit of Halley's Comet should generate new theories, old superstitions and, hopefully, more information. The time of perihelion passage will be Wednesday, 5 February 1986 at 3:49 A.M. EST [later calculations generally agree on a perihelion date of 9 February 1986]. The comet will be observed in the constellation Aries.

1968

1016. Reeves, Paschal. "From Halley's Comet to Prohibition." **Mississippi Quarterly,** 21, No. 4 (1968), 285-90.

About Southern literature and not at all about the comet. However, the article is nonetheless interesting because of its use, in its title, of the comet as a chronological landmark on the literary landscape, evoking the comet's epochal character in the public mind.

1017. Michielsen, H.F. "Rendezvous With Halley's Comet in 1985-1986." **Journal of Spacecraft and Rockets,** 5 (March 1968), 328-34.

As a target of outstanding scientific significance for a space probe, the implications of the comet's predictability on propulsion requirements are developed and numerically evaluated.

1969

1018. Herschel, Sir John. **Herschel at the Cape: Diaries and Correspondence of Sir John Herschel.** David S. Evans, ed. Austin: University of Texas Press, 1969. passim.

Various diary entries for 1835 describe Herschel's efforts and observation of the comet. It is mentioned that the eclipse over Indian Ocean was total on 20 November 1835 and that this is only the second time ever that an eclipse happened concurrent with the apparition of a comet—the other time was 60 A.D.

1019. Jones, D.R.L. and A.J. Meadows. "The Tail Oscillations of Comet Burnham (1960 II) and Comet Halley (1835 III)." **Observatory,** 89 (October 1969), 184-85.

Presents the tabulated elements to support the proposition that the tail of Halley's Comet oscillated in 1835. The probable cause of the oscillation is some form of hydromagnetic perturbation of the comet's plasma sheath which forms its tail.

1020. Shatraw, Milton. "Christmas and the Comet: The Remembrance of a Season on the High Plains." **American West,** 6 (November 1969), 28-32 ff.

The author's reminiscence of seeing the comet in 1910 and his anxiety over reports that the tail of the comet contained poison gas and if earth passed through it all life on earth would perish.

1970

1021. Glasgow, Thurman A. "Father of Space Medicine, An Aerospace Profile: Halley's Comet Launched His Career." **Aerospace Historian,** 17, No. 1 (1970), 6-9.

About Hubertus Strughold who became interested in space because of the 1910 appearance of Halley's Comet.

1022. Classen, J. "Halley's Comet in 1682." **Sky and Telescope,** 39 (February 1970), 102.

The author, an East German astronomer, tells of widespread fear inspired by the comet's 1682 apparition. Facsimile broadside is included.

1971

1023. Brady, J.L. and E. Carpenter. "The Orbit of Halley's Comet and the Apparition of 1986." **Astronomical Journal**, 76 (1971), 728-39.

The motion of the comet from 1682 to 1986 has been investigated and a search ephemeris extending four years on each side of the 1986 perihelion is provided. The last four apparitions have been linked by a continuous integration of 230 years.

1024. Friedlander, A.L. **Halley's Comet Flythrough and Rendezvous Missions Via Solar Electric Propulsion.** Chicago: Illinois Institute of Technology Research Institute, May 1971. 109 pp.

Discusses an electric propulsion system in terms of a Titan launch vehicle and other mission planning factors. Presents trajectory analysis. Available from NTIS, N71-34992. Report no. NASA-CR-122099.

1025. Friedlander, A.L. and W.C. Wells. **Comet Rendezvous Mission Study.** Chicago: Illinois Institute of Technology Research Institute, October 1971. 166 pp.

Four periodic comets with perihelia between 1980 and 1986 (Encke, d'Arrest, Kipff, and Halley) are discussed as potential candidates for the comet rendezvous mission. Available from NTIS, N73-13823. Report no. NASA-CR-129820.

1972

1026. Brady, J.L. "The Motion of Halley's Comet from 837 to 1910." IAU **Symposium**, No. 45 (1972), 155.

Abstract.

1027. Kiang, T. "The Past Orbit of Halley's Comet." **Memoirs of the Royal Astronomical Society,** 76 (1972), 27-66.

Perturbations of the orbit of Halley's Comet over the last 28 revolutions are calculated and a parallel re-examination of Chinese records made.

1028. Brady, Joseph L. "The Effect of a Trans-plutonian Planet on Halley's Comet." **Publications of the Astronomical Association of the Pacific,** 84 (April 1972), 314-22.

An orbit and mass for a hypothetical trans-plutonian planet is determined which reduces the residuals in the time of perihelion passage of Halley's Comet at the seven apparitions from 1910 to 1456 by 93 percent. The effect of this hypothetical planet on the major planets is briefly discussed, and it is shown that the residuals of two similar periodic comets, Olbers and Pons-Brooks are also improved.

1029. Wurm, Karl and Augusto Mammano. "The Motion of Dust and Gas in the Heads of Comets with Type II Tails." **Astrophysics and Space Science**, 18 (October 1972), 491-503.

The Comet Bennet is compared with Comet Halley in 1910; they are related in many respects although Halley's Comet had a lower dust production. Dust particles of tail II are ascribed from the beginning of the expulsion of an electrical charge.

1030. "Halley's Comet and a Hypothetical New Planet." **Sky and Telescope**, 44 (November 1972), 297.

There is some possibility that large telescopes may recover Halley's Comet at the 21st magnitude as early as 1982. The best time of observation in Northern Hemisphere will be at the time of opposition on 18 November 1985. There is a brief discussion of the comet's orbital history.

1973

1031. Visconti, G. and G. Fiocco "Increase of Na Twilight Emission After the Earth's Crossing of the Orbital Planes of Comets Halley and Encke." **Journal of Atmospheric and Terrestrial Physics**, 35 (February 1973), 353-56.

Measurements of sodium twilight emission indicate an increase after the crossing by the earth of the orbital planes of Comets Halley and Encke. Dust particles from the comet enter the atmosphere and fragment into small grains.

1032. Kiang, T. and P.A. Wayman. "The Orbit of Halley's Comet." **Nature**, 241 (23 February 1973), 520-21.

Discusses how a long periodicity could arise in the departure from a theoretical orbit of Halley's Comet through the effect of tangential impulse occurring once per revolution of the comet in its orbit (once every 76 years).

1033. Kiang, T. "The Cause of the Residuals in the Motion of Halley's Comet." **Monthly Notices of the Royal Astronomical Society**, 162 (June 1973), 271-87.

The periodicity on the residuals in perihelion time of comet can be shown to be an inherent property of the three-body configuration of the Sun-Jupiter-Halley.

1034. "Skylab 3 Ready to Study Comet Returning to View."
New York Times, 28 September 1973, 27.

Columns 5-6. About plans to observe Comet Kohoutek.
However, Halley's Comet is mentioned as a reference
point in cometary brightness--Kohoutek, it is thought,
may even outshine Halley's Comet of 1910.

1035. "Note: Science Cruises Becoming Popular." **New York
Times**, 4 November 1973, X, 4.

Columns 5-6. About the upcoming cruise to observe
Comet Kohoutek which may turn out to be brighter than
Halley's Comet as seen in 1910. Again, Halley's Comet
serves as the reference point of visual impact.

1036. Smith, Sherwin D. "The Comet Is Coming." **New York
Times**, 11 November 1973, IV, 118 ff.

About the scheduled appearance of Comet Kohoutek and
its similarity in appearance to that of other comets,
most notably Halley's Comet, is discussed. Also
provided is a reproduction of Halley's Comet from the
Bayeux tapestry and a photograph of Halley's Comet
taken in 1910.

1037. "Comets Used to Be Feared and Unwelcome Visitors."
Wisconsin Now and Then, 20 (December 1973), 2-3.

Reviews the fear and panic experienced in Wisconsin at
the time of the comet's 1910 apparition. Included is a
furniture advertisement for a "seat for comet gazers;"
also given is the recipe for a "comet cocktail": a
snifter of vermouth and a jigger of applejack over
cracked ice." Also reprinted is a 1910 cartoon (see
entry 482).

1038. Stone, Greg. "A Sort of Heavenly Pollywog." **Yankee**,
37 (December 1973), 90-97.

When it was discovered that earth would pass through
tail of the comet in 1910 there was much speculation.
Some predicted a spectacular display of meteors, others
a tremendous explosion and deluge of water. Various
New England newspaper accounts are mentioned.

1039. Weeks, Albert L. "Waiting for Kohoutek." **New York
Times**, 1 December 1973, 33.

Column 2. An op-ed article noting that the scheduled
appearance of Comet Kohoutek in the heavens in December
is causing no particular alarm, in comparison to the
public reaction to the 1910 apparition of Halley's
Comet when there was panic worldwide.

1974

1040. Chase, Virginia. "Once in Maine." **Down East**, 21 (August 1974), 64-67.

A childhood reminiscence of when the author was seven years old. Tells of predictions of the end of the world by fire, and how they affected a child hearing them. Also relates how news of the comet pervaded advertising and the newspapers and how apprehension peaked with the approach of the comet.

1975

1041. Roosen, Robert G. and Brian G. Marsden. "Observing Prospects for Halley's Comet." **Sky and Telescope**, 49 (June 1975), 363-64.

Less than 10 years from now, in late 1984, faint images of Halley's Comet will probably be recorded. The comet will first be visually observable in August 1985 in the morning sky. It is likely that in 1985-86 the comet may disappoint the general public's expectations.

1042. "Halley's Comet: Don't Expect Too Much." **Science News**, 107 (7 June 1975), 367.

Astronomers are warning that the public should not expect a great show from Halley's Comet in 1985 and 1986. The reasons are that the geometry of earth, sun, and comet will be different than in 1910, and also no one knows what the adverse effects of light pollution will be.

1976

1043. Bessel, F.W. **Observations Concerning the Physical Nature of Halley's Comet and Resultant Comments**. Kanner (Leo) Associates: Redwood City, CA., February 1976. 65 pp.

Translated from **Astronomische Nachrichten**, v. 13 (1836), pp. 185-232. Presents two hypotheses designed to explain the behavior of material leaving the nucleus of comet. Available from NTIS, N76-18020/7. Report no. NASA-TT-F-16726.

1977

1044. General Electric Co. **The 200 Watts/Kilogram Solar Array Conceptual Approach Study. Phase 2.** Washington, D.C.: National Aeronautics and Space Administration, 1977. 56 pp.

    Reports on the fabrication, handling, and testing of the solar photoelectric devices for the Halley's Comet mission. Available from NTIS, N77-28580. Report no. NASA-CR-153938.

1045. Yeomans, D.K. "Comet Halley and Nongravitational Forces." in **Comets, Asteroids, Meteorites: 39th AIU Colloquium.** Armand Hubert Delsemme, ed. Toledo, Ohio: University of Toledo, 1977. pp. 61-64.

    Paper presented at the 39th AIU Colloquium, 1976 at the University of Toledo.

1046. Margrave, Thomas E., Jr. "The 1986 Apparition of Halley's Comet." **Physics Teacher,** 15 (February 1977), 110-11.

    Describes a computational astronomy lab exercise appropriate in conjunction with an elementary astronomy or introductory mechanics course as a means of giving practical expression to orbital motion, interplanetary trajectories, Kepler's equation, and the missions to explore Halley's Comet.

1047. Pearson, John F. "Sailing with Halley's Comet--and Other Space Spectaculars for the 1980s." **Popular Mechanics,** 147 (February 1977), 67-71.

    Discusses the solar sail as a spacecraft fuel source. The rendezvous with Halley's Comet may be its first use. Particles in the comet have remained unchanged, in cold storage, as it were, since the birth of the solar system.

1048. Farquhar, R.W. and William H. Wooden, II. **Opportunities for Ballistic Missions to Halley's Comet.** Goddard Space Flight Center: Greenbelt, MD, March 1977. 44 pp.

    Alternative strategies for ballistic missions to Halley's Comet in 1985-86 are described. It is shown that there would be a high scientific return on flyby investment. Available from NTIS, N77-21137/3. Report no. NASA-TM-X-71289.

1049. Wilford, John Noble. "A 1986 Rendezvous Awaiting Halley's Comet." **New York Times,** 2 March 1977, B 1.

Columns 5-6. The scheduled appearance of Halley's Comet in 1986 and the planned scientific observations described. The U.S. is planning a spacecraft to rendezvous with the comet to explore the nature of its nucleus and tail--in order to rendezvous on time the spacecraft would have to embark in late 1981. Conceptual drawings of the spacecraft are included.

1050. "Sailing to Halley's Comet." **Time,** 109 (14 March 1977), 54.

Discusses the inadequacies of traditional rocket fuel sources for a Halley mission, suggesting that the pressure of sunlight be used to power a Halley flight. Gives artist's view of a solar sail during rendezvous.

1051. Watson, Fletcher G. "Get Ready for Halley's Comet." **Physics Teacher,** 15 (May 1977), 260-61.

Letter. Comments on Margrave's February 1977 article (see entry 1046) and describes a three-dimensional orbit model students can build for the comet. Suggests certain times when Halley's Comet might be most spectacular. For northern observers it will be brightest late in April 1986.

1052. Yeomans, Donald K. "Comet Halley--the Orbital Motion." **Astronomical Journal,** 82 (July 1977), 435-40.

The orbital motion of the Comet Halley is investigated over the period 837-2061 A.D. For the 1986 return, viewing conditions are outlined for the comet and the related Orionid and (eta) Aquarid meteor showers.

1053. National Aeronautics and Space Administration. **Scientific Rationale and Strategies for a First Comet Mission. Report of the Comet Halley Science Working Group. An Executive Summary.** National Aeronautics and Space Administration: Washington, D.C., July 1977. 11 pp.

Presents the justification, scientific objectives, instrumentation, and strategy for a first comet mission are discussed. Support research to be done by NASA is also discussed. Is available from NTIS, N77-29044/3. Report no. NASA-TM-78420-PT-2. See entry 1054 for unabridged document.

1054. National Aeronautics and Space Administration. **Scientific Rationale and Strategies for a First Comet Mission. Report of the Comet Halley Science Working Group.** National Aeronautics and Space Administration: Washington, D.C., July 1977. 92 pp.

Conference proceedings. The science objectives of a first comet mission are reviewed and related to what is known or can be expected to be learned. Instruments and science objectives are defined. Available from NTIS, N77-29043/5. Report no. NASA-TM-78420. See entry 1053 for summary document.

1055. West, J. **Ion Propulsion Module (IPM) Technology Readiness Assessment: 1985 Halley's Comet Rendezvous Mission.** Jet Propulsion Lab: Pasadena, CA., 15 July 1977. 19 pp.

An assessment of the risk of utilizing ion propulsion to perform a rendezvous mission with Halley's Comet in 1985 is presented. Recommendations are considered for reducing the identified risks to lowest possible level. Is available from NTIS, 78-18120/3. Report no. NASA-CR-155308.

1056. Hughes, David W. "Visiting Halley's Comet." **Nature,** 268 (11 August 1977), 486-88.

Discusses the comet missions that are planned and their goals: to image the nucleus, to find its size, shape, rotation period and color, to determine the nature of the multiple condensations seen in 1910, and to determine the abundance and spatial distribution of nucleus molecules.

1057. Rayl, G.J. et al. **Solar Array Conceptual Design for the Halley's Comet Ion Drive Mission, Phase 2 (Final Report).** Jet Propulsion Lab: Pasadena, CA., 31 August 1977. 87 pp.

Conceptual design studies were performed toward a high power, ultralightweight solar array, compatible with the requirements for the Halley's Comet Ion Drive Mission. Available from NTIS, N78-16438/1. Report no. NASA-CR-155594 or DOC-77SDS4243.

1058. "Rendezvous With Halley's Comet." **Mechanical Engineering,** 99 (September 1977), 48.

Included is an artist's conceptual drawing of the solar electric propulsion system proposed by NASA for the Halley's Comet flyby. A launch date between March and June 1982 is proposed to send the craft on a three-and-a-half year 600 million mile journey.

1059. Norman, Colin. "No Long Encounter with Halley's Comet." **Nature,** 269 (22 September 1977), 280.

NASA has reluctantly abandoned the plan to conduct a close-range study of Halley's Comet. The plan would have cost 500-600 million dollars and there is simply not enough money in the budget. NASA, however, still may launch a craft to fly through the tail without a special propulsion system.

1060. "Halley's Comet." **Sky and Telescope**, 54 (November 1977), 363-64.

Briefly discusses some current literature devoted to the comet.

1061. Fearn, D.G. "Missions to Halley's Comet." **Nature**, 270 (10 November 1977), 97.

Letter. Clarifies D.W. Hughes's article (see entry 1056) to the effect that there are indeed ion-type thrusters which are highly effective, and presently ready for use on a comet mission such as that being proposed for Halley's Comet.

1978

1062. "Comet Halley Photographed May 19, 1910 at Lowell Observatory." [Tucson?]: Kitt Peak National Observatory, 1978.

Poster. "Computer enhancement and color produced by the picture processing lab of the Kitt Peak National Observatory." Copies can be ordered from KPNO and suppliers such as the Hansen Planetarium in Salt Lake City, and the Astronomical Society of the Pacific in San Francisco.

1063. Arnold, Armin. "Halley's Comet and Jakob van Hoddis' Poem 'Weltende'." in **German Expressionism**. Victor Lange. New York: Griffon House, 1978. pp. 47-58.

Some literary critics and historians have traced the origin of the German Expressionist movement to this poem, unaware that it was inspired by the Halley's Comet panic of 1910.

1064. West, J.L. "Ion Drive Technology Readiness for 1985 Halley Comet Rendezvous Mission." **Astronautics and Aeronautics**, 16, No. 12 (1978), PB 18.

This paper was originally given at AIAA/DGLR 13th International Electric Propulsion Conference, San Diego, California, 25-27 April 1978. It can be ordered from AIAA as paper no. 78-641.

1065. Wood, L.J. and S.L. Hast. "Navigation Accuracy Analysis for the Halley Flyby Phase of a Dual Comet Mission Using Ion Drive." NY: American Institute of Aeronautics and Astronautics, 1978.

The paper was given at AIAA 18th Aerospace Sciences meeting, Pasadena, California, 14-16 January 1980. The paper can be ordered from the AIAA as Technical Paper no. AIAA-80-0115.

1066. "Solar Array for Halley's Comet Mission." **Space World**, O-1-169 (January 1978), 27.

The solar array is a vital element in an ion drive propulsion system being considered because it can provide a continuous low thrust to catch and follow the comet. Presents an artist's rendering of an ion engine driven spacecraft on its way to rendezvous with Halley's Comet.

1067. Horsewood, J.L. and F.I. Mann. **SEP Encke-87 and Halley Rendezvous Studies and Improved S/C Model Implementation in HILTOP.** Seabrook, MD: Business and Technological Systems, Inc., February 1978. 29 pp.

Studies were conducted to determine the performance requirements for projected state-of-the art spacecrafts for a Halley's Comet rendezvous. Available from NTIS, N78-25105/5. Report no. NASA-CR-135414.

1068. Hughes, David W. "Here Comes Halley's Comet!" **Sciences**, 18 (February 1978), 6-9, 31.

Contains suggestions for a space mission to intercept the comet. The cover of the issue offers a color reproduction of the Bayeux tapestry representation of the comet's 1066 appearance.

1069. Covault, Craig, "NASA Plans New Space Missions." **Aviation Week and Space Technology**, 108 (17 April 1978), 14-16.

A Halley mission would involve a Space Shuttle launch in August 1985. Discusses other mission options and strategies. However, Halley's Comet is the only bright comet for which a return to the vicinity of the earth can be predicted accurately and is therefore a desirable target and scientific challenge. The data compiled shows the comet's major features to be nuclear region and coma, tail, and dust (a brief explanation of each is given). An artists rendition of the spacecraft encountering the comet is provided.

1070. "Scientists Plan Trip to Halley's Comet." **New Scientist**, 78 (18 May 1978), 429.

Announces that the European Space Agency (ESA) will send a comet mission to Halley's Comet costing about a half-billion dollars. NASA was originally unconvinced of the value of a comet mission. The likelihood is that ESA's revised comet flyby plan will be approved. Included is an artist's conception of the flyby spacecraft.

7. Artist's rendition of a proposed NASA spacecraft intercepting Halley's Comet 93 million miles from earth. Note the deployment of a probe toward the body of the comet. See entry 1069. Reprinted by courtesy of *Aviation Week and Space Technology* and the National Aeronautics and Space Administration.

1071. Thornton, Catherine L. and Robert A. Jacobson. "Navigation Capability for an Ion Drive Rendezvous with Halley's Comet." **Journal of Astronautical Sciences,** 26 (July 1978), 197-210.

   Navigation accuracies are presented for a 1985 rendezvous with Halley's Comet using an ion propulsion system. Individual error sources are examined in order to determine their contribution to final delivery errors.

1072. Jacobson, R.A. and C.L. Thornton. "Elements of Solar Sail Navigation With Application to Halley's Comet Rendezvous." **Journal of Guidance and Control,** 1 (September 1978), 365-71.

   The problem of interplanetary navigation of a solar sail spacecraft is examined and found to be analogous to that of solar electric craft. An evaluation of the strategy for a Halley's Comet rendezvous is made.

1073. Chartrand, Mark R. "Comet Chasing: Space." **Omni,** 1 (October 1978), 24-25.

   About NASA plans to launch a Halley/Temple 2 mission in 1985. In times past comets were thought to be the finger of God pointing at the earth, thereby bringing disaster (dis-aster means bad star). Support for scientific research is seen to be an indication of a society's maturity.

1979

1074. Atkins, Kenneth L. "Ion Propulsion and Comet Halley Rendezvous." in **Space Missions to Comets.** Marcia Neugebauer et al., ed. Washington, D.C.: National Aeronautics and Space Administration, 1979. pp. 197-217.

   Presented at a conference sponsored by the NASA Office of Space Science at the Goddard Space Center, October 1977. Describes the ion engine (looking like a coffee can 15-inches high) and offers that it is a viable alternative to the considerable energy problems inherent in a comet mission of the type under discussion; illustrated with various diagrams and drawing. Analogy used of "Moby Comet," this great white ghost in the sky, with the spacecraft as our ship, taking us out to the comet so that our harpoon can sample it. SUDOC classification NAS1.55:2089.

1075. Delsemme, A.H. "Scientific Returns from a Program of Space Missions to Comets." in **Space Missions to Comets.** Marcia Neugebauer et al., ed. Washington, D.C.: National Aeronautics and Space Administration, 1979. pp. 139-78.

Presented at a conference sponsored by the NASA Office of Space Science and held at Goddard Space Flight Center, October 1977. After enumerating what could conceivably be expected to be learned from cometary missions, Delsemme marks Halley's Comet as, by far, the best target for a first comet mission. It has a fairly reliable brightness and orbital behavior and has a gas production rate two orders of magnitude greater than any other comet whose passage can be reliably predicted before 2010. For this reason, more accurate and sensitive measurements of its chemical composition are possible. It is the only reliable comet to display the full range of cometary phenomena. SUDOC classification NAS1.55:2089.

1076. Marsden, Brian G. "Comet Halley and History." in **Space Missions to Comets**. Marcia Neugebauer et al., ed. Washington, D.C.: National Aeronautics and Space Administration, 1979. pp. 179-96.

Presented at a conference sponsored by the NASA Office of Space Science and held at Goddard Space Flight Center, October 1977. Presents a delightful compendium of astronomical history and events associated with comet's various apparitions. SUDOC classification NAS1.55:2089. See also entry 1158.

1077. Wallis, Max K. "Probing Energetic Particles in Comet Halley." **Geophysical Journal of the Royal Astronomical Society**, 57, No. 1 (1979), 303.

Abstract. Of the various phenomena in the cometary coma involving energetic particles--bow shock, ion acceleration, plasma envelope and ray formation, contact disconuity, etc., only limited examination would be possible during the proposed fast and distant flyby.

1078. Atkins, Kenneth L. **Mission Summary: Halley Flyby/Tempel-2 Rendezvous.** Jet Propulsion Lab: Pasadena, CA., 15 January 1979. 13 pp.

A unique dual-comet flight opportunity exists in mid-1985 that includes the flyby of a large and active comet en route to rendezvous with a second comet, Tempel 2. The mission should be launched in July 1985. Available from NTIS, 79-15133/8. Report no. NASA-CR-158072 or JPL-PUB-79-4.

1079. Broughton, R. Peter. "Visibility of Halley's Comet." **Journal of the Royal Astronomical Society of Canada**, 73 (February 1979), 24-36.

When first seen during each apparition throughout its recorded history the comet has been of the same intrinsic brightness. However, the comet and the earth will be on opposite sides of the sun in February 1986 making circumstances the worst for observation in the last 2000 years.

1080. "European Group Agrees to Comet Mission Role." **Aviation Week and Space Technology**, 110 (26 February 1979), 20-21.

The European Space Agency has agreed to a 15% participation in NASA's proposed comet mission in an attempt to avoid a one-year delay because of budget restrictions.

1081. Chang, Y.C. "Halley's Comet: Tendencies in Its Orbital Evolution and Its Ancient History." **Chinese Astronomy**, 3 (March 1979), 120-31.

The method used to study the orbital evolution of Halley's Comet during the past three thousand years is discussed based on the motion of the comet and the perturbations of the nine major planets, in conjunction with historical records.

1082. Brin, G.D. and G.D. Mendis. "Dust Release and Mantle Development in Comets." **Astrophysical Journal**, 229 (1 April 1979), 402-08.

On the assumption that the nucleus of a comet consists of a uniform mixture of dust and ice, a model is developed and applied specifically to the "gassy" periodic comet P/Halley.

1083. Olson, Roberta J.M. "Giotto's Portrait of Halley's Comet." **Scientific American**, 240 (May 1910), 160-68.

A blazing comet represents the star of Bethlehem in one of Giotto's famous ("The Adoration of the Magi") Arena Chapel frescoes in Padua. It is a naturalistic portrait of Halley's Comet during its apparition of 1301.

1084. "Halley's Comet: Past, Present and Future." **Sky and Telescope**, 58 (September 1979), 232.

Interest in the comet is growing even though the coming apparition will be relatively unfavorable.

1085. Hughes, David W. "The Micrometeoroid Hazard to a Space Probe in the Vicinity of the Nucleus of Halley's Comet." in **The Comet Halley Micrometeoroid Workshop.** Washington, D.C: National Aeronautics and Space Administration, October 1979. pp. 51-56.

The micrometeoroid impact rate on the Halley probe is calculated as a function of the flyby distance The number of meteoroid particles in specific log mass ranges which will impact on a unit area of the leading surface of the probe are given. When producing a model of the environment in the inner coma of Halley's Comet two of the important parameters required in the calculation are the mass and the size of the central nucleus, the fount of the comet's activity. Unfortunately, mass is on of the most awkward quantities to measure in an astronomical context. Suggests that analysis of mass and diameter of the comet indicates that $1.4 \times 10/17$ g and 6.2 km are reasonable values. Presented at the ESA International Workshop, Noordwijk, Netherlands, 18-19 April 1979. Available from NTIS, 80-22200/3 or from ESA as Special Publication no. 153.

1086. Sekanina, Z. "Expected Characteristics of Large Dust Particles in Periodic Comet Halley." in **Comet Halley Micrometeoroid Workshop.** Washington, D.C.: National Aeronautics and Space Administration, October 1979. pp. 25-34.

Observational evidence in presented for dust particles heavier than $10/-8$ gram expelled from Periodic Comet Halley. Dust emitted around 1.5 AU preperihelion, the point in orbit where a probe would fly through the comet in 1985, should have contributed to an anomalous tail that could have been observed in mid to late april 1910 and again between 21 and 27 May 1910. In 1835 conditions for an anomalous tail were satisfied after conjunction with the sun in early December. Its observation by J. Herschel in late January 1836 is used to derive the production of dust vs. time and the expected number of large-particle impacts on the protective shield of the 1985 flythrough probe. Presented at the ESA International Workshop, Noordwijk, Netherlands, 18-19 April 1979. Available from NTIS, 80-22200/3 or from ESA as Special Publication no. 153.

1087. Hughes, David W. "Halley's Comet in Ancient History." **Nature,** 281 (11 October 1979), 428.

Discusses research done in Chinese astronomical records that makes the 1985/86 apparition the 41st recorded apparition, instead of the 30th as previously was thought.

1088. Large, Arlen J. "The Ion Engine's Glamorous Mission Is Chasing Comets." **Wall Street Journal,** 11 October 1979, 1, 41.

Column 4. An ion powered device may power the spacecraft that observes Halley's Comet--if NASA gets adequate funding. The ion engine requires electricity which is furnished by solar cells facing the sun. It will accelerate the spacecraft to 31,540 miles per hour as it flys by the comet.

1089. "Chasing Comets." **New York Times**, 20 November 1979,
C 2.

Column 1. Scientists were asked last week to propose
experiments for a planned 1985 international mission to
explore Halley's Comet and Comet Tempel 2. This was
done even though the mission has not yet been approved
by Congress.

1090. "ESA-NASA to Launch Halley-Tempel-2 Comet Mission."
Nature, 282 (22 November 1979), 352.

The mission will last four years. NASA is to be
responsible for the launching and mission operations
while the ESA will be responsible for the Halley's
Comet probe system.

1091. "OMB Kills Halley-Tempel 2 Mission." **Aviation Week
and Space Technology**, 111 (26 November 1979), 20.

The spacecraft mission that would send a probe to
Halley's Comet and then rendezvous with Comet Tempel 2,
and possibly land on its surface, has been killed by
the Office of Management and Budget action on the NASA
FY 1981 budget request.

1092. "Discussion." **Scientific American**, 240 (December
1979), 13-14.

Based on his reading of Gibbon's **The History of the
Decline and Fall of the Roman Empire** Samuel F. Howard
queries as to whether Emperor Justinian's comet of A.D.
530 could have been Halley's Comet (in reference to
Olson's article in **Scientific American**, May 1979; see
entry 1083). Olson responds that it was not Halley's
Comet and that the confusion probably arises from the
difference in Gregorian and Julian calendars.

1093. "Ride a Comet's Tail on Wings of the Sun." **Space
World**, P-10-192 (December 1979), 4-8.

Discusses the Halley's Comet rendezvous mission and ion
engines. With this apparition man will go out and meet
the comet and learn more in that brief encounter than
since the beginning of recorded history. Reprinted
from **Vectors**, a publication of Hughes Aircraft Company.

1094. Covault, Craig. "Fund Cut Forces Comet Strategy
Shift." **Aviation Week and Space Technology**, 111 (3 December
1979), 61-63.

NASA looks at a new rendezvous strategy involving
limited operations and higher spacecraft costs. The
plan is controversial. Includes illustrations of the
geometry of the comet's approach to sun and earth as
well as two spacecraft flybys.

1095. European Space Agency. **The International Comet
Mission: Report   on the Phase A Study.** Washington, D.C.:
National Aeronautics and Space Administration, 5 December
1979. 68 pp.

The scientific objectives, mission analysis, system
design and management, and time schedule for the
Halley/Tempel 2 Comet Mission are outlined. Current
knowledge about comets is summarized. Available from
NTIS, N80-22385/2. Report no. ESA-SCI(79)11.

1096.   "First Contracts Given for a Halley's Comet
Mission." **New York Times,** 12 December 1979, 25.

Column 1. NASA awards contracts to Boeing and Lockheed
to develop a spacecraft with new solar rocket engines
for a rendezvous with Halley's Comet. NASA emphasizes
that it still needs White House and congressional
approval.

1097. "Comet Mission Hits a Snag." **Science News,** 116 (15
December 1979), 407.

Scarcely a month ago NASA asked scientists to propose
experiments for a spacecraft that would sweep past
Halley's Comet. Now the entire mission may be scrapped
because of close-cut budgets that would not fund the
Solar Electric Propulsion System.

1098. "Tailing a Comet." **Time,** 114 (24 December 1979), 70.

Announces NASA's awarding of contracts for studies of a
new rocket to Boeing and Lockheed. Plans for a comet
mission are contingent on congressional funding.

1099. Wilford, John Noble. "Space Shuttle Problems Cut
Support for Other Missions." **New York Times,** 31 December
1979, 10.

Column 1. The effect of eroding congressional support
for the space shuttle on the planned 1985 Halley probe
is discussed. NASA had planned to fire the comet probe
into the head of the nucleus of Halley's Comet and then
proceed on to Comet Tempel 2--now NASA might have to
settle just for a Halley's Comet probe.

1980

1100. Akiba, Ryojiro et al.   "Orbital Design and
Technological Feasibility of Halley Mission." **Acta
Astronautica,** 7 (1980), 797-805.

Missions to Halley's Comet in 1985-86 are discussed.
They include a pre-perihelion encounter, post

211

perihelion encounter, utilization of a Venus swing-by, and a mission around the sun before the actual comet encounter. The technological feasibility of a small spacecraft is discussed.

1101. Boain, R.J. "A Ballistic Mission to Flyby Comet Halley." New York: American Institute of Aeronautics and Aerodynamics, 1980.

This paper was given at AIAA/AAS Astrodynamics Conference, Danvers, Massachusetts, 11-13 August 1980. The paper may be ordered as Technical Paper no. AIAA-80-1686.

1102. Brandt, J.C. et al., eds. **The International Halley Watch: Report of the Science Working Group.** Washington, D.C.: National Aeronautics and Space Administration, 1980. 72 pp.

Plans are presented for coordinating professional and amateur observations over a broad range of studies. The appendix includes an ephemeris at 5-day intervals. Available as NASA Technical memorandum 82181, SUDOC classification NAS1.55:82181.

1103. Hughes, David. "Mission to the Comets." **New Scientist,** 85 (1980), 66-69.

A brief report on plans for a spacecraft mission to comets Halley and Tempel-2.

1104. Reinhard, R. and D. Dale. "Giotto--A Mission to Halley's Comet." **European Space Agency Bulletin,** No. 24 (1980), 6-11.

Halley's Comet's next apparition is an opportunity for scientific study not to be missed. The spacecraft for the GIOTTO project is described in regard to its major features. The principal goals for the spacecraft's scientific payload include imaging of the comet's nucleus, and measurements of the elemental and isotopic composition of the cometary gases and dust. Diagrams the spacecraft and the locations of the various instruments. Also diagrammed is the geometry of the Halley encounter.

1105. Farquhar, R.W. and W.H. Wooden, II. **Halley's Comet 1985-86: A Unique Opportunity for Space Exploration.** Greenbelt, MD: Goddard Space Flight Center, January 1980. 15 pp.

A coordinated program to explore Halley's Comet in 1985-86 is proposed. Describes the interaction of spacecraft with ground based stations. Available from NTIS, 80-30343/1. Report no. NASA-TM-80633.

1106. "A Sad Tale of Two Comets." **Astronomy,** 8 (February 1980), 16-23.

Of all solar system bodies known, none is more pristine than a comet--a perfect record of earliest time preserved. Discusses the nature and benefits of the various space probes directed at Halley's Comet and Tempel 2. Provides a comet data table.

1107. Cameron, A.G. "Statement/Discussion." **1981 NASA Authorization, Vol. IV. House Committee on Science and Technology serial No. 96-111.** Washington, D.C.: U.S. Government Printing Office, February 1980. pp. 1467-72.

Cameron, Chairman of the Space Science Board of the National Academy of Sciences, testifies that at this time the orbit of Halley's Comet is such that a rendezvous is impossible--primarily because of the delay in funding the necessary solar electric propulsion system (SEP) for the rendezvous spacecraft. SUDOC classification Y4.Sci2:96/111.

1108. Mutch, Thomas A. "Statement/Discussion." **1981 NASA Authorization, Vol. IV. House Committee on Science and Technolgy serial No. 96-111.** Washington, DC: U.S. Government Printing Office, February 1980. pp. 1381-1466.

Mutch is Associate Administrator for Space Science at NASA. He testifies in support of NASA program requests for a Halley mission. The importance of the Halley mission is seen by the lack of emphasis given to it within the context of the total NASA program proposals. SUDOC classification Y4.Sci2:96/111.

1109. Mutch, Thomas A. "Statement/Discussion." **NASA Authorization for FY81. Part 2. Senate Committee on Commerce, Science, and Transportation serial No. 96-62.** Washington, DC: U.S. Government Printing Office, February 1980. pp. 1151-1244.

Mutch is Associate Administrator for Space Science at NASA. He very briefly discusses the prospects for observation of Halley's Comet. In fact, this testimony is more notable for the lack of attention paid to a Halley mission than any thing said about it. SUDOC classification Y4.C73/7:96-62/pt.2.

1110. Veverka, Joseph. "Statement/Discussion." **1981 NASA Authorization, Vol. IV. House Committee on Science and Technology serial No. 96-111.** Washington, DC: U.S. Government Printing Office, February 1980. pp. 1473-87.

Veverka is a planetary scientist at Cornell University. He presents the Space Science Board's recommendations emphasizing comet observation. Veverka points out that unless development of the solar electric propulsion system is begun very soon a mission to Halley's Comet will be virtually impossible. SUDOC classification Y4.Sci2:96/111.

1111. "Comets In a Storm: Tight Money for Space." **Science News**, 117 (16 February 1980), 101-02.

Considerable furor has arisen out of concern among some scientists and others over the future of the proposed Halley mission. Points out that, Thomas Mutch, the Associate Director of NASA in 33 pages of congressional testimony devoted only 9 sentences to the comet mission (see entries 1107 and 1108).

1112. Sagan, Carl. "Statement." **NASA Authorization for Fiscal Year 1981: Hearings before the Senate Subcommittee on Science, Technology, and Space, serial no. 96-62.** Washington, D.C.: U.S. Government Printing Office, 27 February 1980, pp. 1638-40.

Speaks forcefully and eloquently in favor of a Halley's Comet mission because it provides the opportunity for detailed in situ studies. "Unfortunately we cannot rendezvous at all with Comet Halley, which is a longer periodic comet on a retrograde orbit. Nonetheless, Halley is an important scientific and popular occasion; it is prudent for NASA to respond to its apparition later this decade. It is the only relatively young and large comet that we will pass in this century; it offers an important and exciting opportunity both for the first scientific investigation of a comet by spacecraft and for maintaining and stimulating public interest." Goes on to recommend U.S. organization of, support for, and participation in an "international Halley watch" with the realization that "the NASA program does not exist for scientists only. And although its planning should be done in a scientifically conscientious and responsible manner, the interest of the public in this most important celestial event should not be overlooked. The comet will not return for another 76 years." Points out that with a high-quality imaging system on board a spacecraft, Halley's Comet could be brought into millions of homes throughout the world. SUDOC classification is Y4.C73/7:96-62/pt.3.

1113. "Coming of Halley: The World Readies." **Science News**, 117 (15 March 1980), 167-68.

NASA plans for studying Halley's Comet have been in trouble for years. "But Halley has a unique allure. Besides being the only comet people can name, it has been described as the 'most scientifically valuable single comet in the sky'."

1114. Fisher, Arthur. "Close Encounter With A Comet."

Popular Science, 216 (April 1980), 15-16.

Reviews the preparation for the space probes of Halley's Comet and the request that has gone out to the scientific community for submission of proposed experiments.

1115. Franklin, Kenneth L. "Here Comes the Heaven's Most Notorious 'Dirty Snowball'." **Science Digest**, 87 (April 1980), 30-33.

"Ask anyone--even a stranger--to name a comet, and the chances are high that you will hear: 'Halley's Comet.'" During its 1985-86 apparition the comet will be disappointing for two reasons: 1. Poor timing and placement; and 2. The comet is wearing out.

1116. **Giotto Comet Halley Mission, Phase a Study.** Volume 1: **Spacecraft Design (Final Report).** Washington, D.C.: National Aeronautics and Space Administration, May 1980. 125 pp.

A study done by British Aerospace Dynamics Group, Bristol (England) in cooperation with Dornier-Werke G.M.B.H., Friedrichshafen, Germany. A spacecraft for a post perihelion encounter with Halley's Comet was designed. Available from NTIS, N82-19313.

1117. "Scientists Keep Watch for Comet's Leftovers." **New York Times**, 4 May 1980, I, 27.

Column 6. Astronomers kept watch this morning for shooting stars in the southeastern sky that they believe to be the "leftover" particles from Halley's Comet. The comet will be seen again on 25 October 1985 and on 8 May 1986.

1118. European Space Agency. **Giotto: Comet Halley Flyby. Report on the Phase a Study.** Washington, D.C.: National Aeronautics and Space Administration, 14 May 1980. 87 pp.

A study of the type of rocket and satellite to be used to intercept the nucleus of Halley's Comet. The present knowledge of comets is reviewed. Mission description including spacecraft and system configuration is presented. Available from NTIS N80-30346/4. Report no. ESA-SCI(80)-4.

1119. Redfearn, Judy. "Meeting Halley." **Nature**, 285 (26 June 1980), 609.

The ESA plan for a Halley's Comet mission is reviewed. The planned flight would pass within 1000 km. of the nucleus of the tail side of the comet. Such a pass presents a high risk of damage from dust particles or micrometeorites. A diagram shows the orbit of the

215

comet and the planned track of the ESA spacecraft.

1120. National Aeronautics and Space Administration. **The International Halley Watch.** Washington, D.C.: National Aeronautics and Space Administration, July 1980. 80 pp.

In preparation for the 1985-86 apparition the International Halley Watch has initiated a comprehensive program. Plans and ground based observing nets are discussed in detail. Also presented is a synoptic history of comet. Available from NTIS, N80-32310/8. Report no. NASA-TM-82181.

1121. Williams, Gurney, III. "Will Halley's Comet Fizzle?" **Popular Mechanics,** 154 (July 1980), 14.

Communicates that Professor Miriam Jaffe of Purdue University suggests that the relative positions of the comet, earth, and sun will reduce the comet's glitter. When Halley's Comet is closest to the sun on 9 February 1986 the earth will be far away on the opposite side of the sun.

1122. Covault, Craig. "European Space Unit to Get Proposal on Comet Flyby." **Aviation Week and Space Technology,** 113 (7 July 1980), 38-39 ff.

Presents excellent diagrams of the Halley's Comet flyby spacecraft and of the principal features of Halley's Comet. A ten-year effort to define a mission to a comet approaches culmination with a U.S. proposal. Discusses the planning for the various comet missions.

1123. "Mission to Halley Is In the Cards Again." **New Scientist,** 87 (10 July 1980), 95.

Europe will probably send a spacecraft to Halley's Comet after all--the ESA Science Advisory Board has unanimously recommended that the flight be undertaken.

1124. "ESA Opts for Own Halley Comet Flyby." **Aviation Week and Space Technology,** 113 (14 July 1980), 27.

ESA's science program committee voted 10 to 1 in favor of a sole European mission. ESA is willing to reconsider this decision if NASA is willing to provide the launcher.

1125. "Europeans to Probe Halley's Comet Without U.S. Help." **Wall Street Journal,** 15 July 1980, II, 31.

Column 6. Europe will launch its own satellite in 1985 because of cuts in the American space budget. ESA will spend $125 million on what is its most ambitious project to date. The probe will push to within 300-1200 miles of comet.

1126. "ESA Comet-bound With or Without NASA." **Science News**, 118 (19 July 1980), 38-39.

ESA will go it alone if NASA cannot get the funding for a Halley's Comet mission. The flyby will happen at a relative velocity of 68 kilometers per second, so all mission observations will be made in a single 4 hour period. The data will be immediately radioed to earth.

1127. Houpis, Harry L.F. and D.A. Mendis. "Physicochemical and Dynamical Processes in Cometary Ionosphere. I. The Basic Flow Profile." **Astrophysical Journal**, 239 (1 August 1980), 1107-18.

There is an indication that the nuclear distance of the stagnation point is approximately-greater-than four times $10/(3)$ km., for a medium-bright comet like P/Halley at a heliocentric distance of 1 AU, which is well outside the region of a collisional coupling between cometary ions and neutrals.

1128. Covault, Craig. "NASA Halley Bid Opposed, Supported." **Aviation Week and Space Technology**, 113 (20 October 1980), 107-08.

Discusses the behind the scenes politics of moving toward a decision for U.S. participation or nonparticipation in the European Space Agency's GIOTTO Halley's Comet intercept program.

1129. Jet Propulsion Lab. **Comet Science Working Group Report on Halley Intercept Mission.** Washington, D.C.: National Aeronautics and Space Administration, November 1980. 22 pp.

The Halley Intercept Mission is described and the scientific benefits expected from the program are defined. Specific recommendations are made concerning project implementation and spacecraft requirements. Available from NTIS, N81-20100/6. Report no. NASA-TM-82386.

1130. Sullivan, Walter. "Comet Missions Due, But U.S. Project Is in Doubt." **New York Times**, 3 November 1980, 13.

Columns 1-6. Japan, the Soviet Union, and the European Space Agency (ESA) are planning missions to intercept Halley's Comet; but the U.S. mission has been given a low priority in the Congress and therefore its future remains in doubt.

1131. "Will France Be First to Halley's Comet.?" **New Scientist**, 88 (6 November 1980), 353.

The French Space Agency is hoping to steal a march on its big brother, the European Space Agency, by teaming up with the Soviets to send a satellite to rendezvous with Halley's Comet.

1132. "Soviets Revise 1984 Unmanned Mission Plan to Include 1986 Flyby of Halley's Comet." **Aviation Week and Space Technology**, 113 (10 November 1980), 18-19.

Describes the Soviet strategy for coordinating Venus and Halley's Comet flybys. Discusses the French contribution to the project.

1133. Oppenheimer, Michael and Leonie Haimson. "The Comet Syndrome." **Natural History**, 89 (December 1980), 54-61.

"Human reactions to comets range from hysteria to frivolity." Approximately half of the article deals with Halley's Comet and our reactions to it through the course of history.

1134. Covault, Craig. "Mission to Halley." **Aviation Week and Space Technology**, 113 (15 December 1980), 11.

A decade of visionless NASA budgeting has crippled instrument development for cometary exploration. A flight to Halley's Comet is an explorer's dream, scientifically valuable, adventuresome, and exciting. It is a mission with the potential for public appeal.

1135. Dickson, David. "NASA to Go?" **Nature**, 288 (18 December 1980), 634-35.

Hopes that the lack of support from the Carter administration will not carry over to the Reagan administration. Relates European interest in the U.S. Halley mission prospects in light of their own Halley mission (GIOTTO) plans.

1136. Cowen, Robert C. "Can the U.S. Keep Its Lead among the Planets?" **Christian Science Monitor**, 31 December 1980, 12.

Column 1. For U.S. planetary scientists the mention of Halley's Comet inspires gloom vis a vis a missed opportunity. It symbolizes the loss of impetus that the U.S. has let befall its planetary space exploration program.

1981

1137. Anderson, Norman D. and Walter R. Brown. **Halley's Comet**. New York: Dodd, 1981. 78 pp.

Two chapters are devoted to matching historical disasters with comet appearances. Black and white photos. This book is specifically geared to the

9-to-15-year-old age group, is non-technical in its presentation of scientific astronomical facts and non-scientific in its presentation of astronomical history.

1138. Batchelor, John Calvin. **The Further Adventures of Halley's Comet.** New York: Condon and Lattes, 1981.

Novel. A swashbuckling adventure set in 1986 when Halley's Comet is making its next pass by the earth. Replete with pseudo-historical flashbacks to the last four apparitions of Halley's Comet.

1139. Calder, Nigel. **The Comet Is Coming! The Feverish Legacy of Mr. Halley.** New York: Viking, 1981. 160 pp.

Halley's Comet serves as the center piece for this study of comets. The book is directed at a general audience, and provides an interesting mix of astronomical history from the earliest time through to the present planned space missions. There is also ample comet lore and superstition about comets in general and Halley's Comet in particular. Done in a style that is wryly entertaining--refers to the comet as "orbiting trash" and "sky pollution."

1140. Divine, Neil. "Numerical Models for Halley Dust Environments." in **The Comet Halley Dust and Gas Environment.** Paris: European Space Agency, 1981, pp. 25-30.

The algebra, parameter values, and sample results of a Reference Model for the dust of Comet P/Halley are presented. The model employs simplified treatments of the icy conglomerate nucleus and the hydrodynamic outflow of gas and dust. Novel features of the model include a gas production rate newly derived from observations of Halley's visual magnitude in 1910, gas and dust flux non-uniformly distributed over the nuclear surface, a dust distribution function consistent with large particle inferences from the comet antitails, dust density variable with mass, and a reasonable approximation to the distribution of particles near the dust envelope. Presented at the Joint NASA/ESA Working Group Meeting, Heidelberg, 26 and 27 August 1981. Available as ESA Special Publication no. 174.

1141. Divine, Neil. "A Simple Radiation Model of Cometary Dust for P/Halley." in **The Comet Halley Dust and Gas Environment.** Paris: European Space agency, 1981, pp. 47-53.

The model includes absorption and single scattering of sunlight, and infrared emission, for the dust and for the nucleus. It employs dust particle photometric

219

properties which are independent of wave-length, particle size, and position within the coma, separately in the visible and infrared regions of the spectrum. Presented at the Joint NASA/ESA Working Group Meeting at Heidelberg, 26 and 27 August 1981. Available as ESA Special Publication no. 174.

1142. Hanner, Martha S. "Effect of the Size Distribution of the Thermal Emission from Cometary Dust." in **The Comet Halley Dust and Gas Environment.** Paris: European Space Agency, 1981. pp. 67-75.

The thermal emission from cometary dust is analyzed in terms of two size distribution functions. The resulting thermal emission spectra are illustrated for heliocentric distances 0.15 AU-1.6 AU. Presented at the Joint NASA/ESA Working Group Meeting at Heidelberg, 26-27 August 1981. It is available as ESA Special Publication no. 174.

1143. Newburn, R.L. and R. Reinhard. "The Derivation of Halley Parameters from Observations." in **The Comet Halley Dust and Gas Environment.** Paris: European Space Agency, 1981. pp. 19-24.

A set of nominal value model parameters for the nucleus, the dust and gas for P/Halley at 0.9 AU post-perihelion is derived from the light curve and spectra of Halley and by modeling the effect of the nongravitational forces. A 'flow diagram' shows how the parameter values are derived and to what extent these derived values are interdependent. Present at the NASA/ESA Joint Working Group Meeting at Heidelberg, 26-27 August 1981. Available as ESA Special Publication no. 174.

1144. Newburn, Ray L. "A Semi-Empirical Photometric Theory of Cometary Gas and Dust Production: Application to P/Halley's Gas Production Rates." in **The Comet Halley Dust and Gas Environment.** Paris: European Space Agency, 1981. pp. 3-18.

The semi-empirical photometric theory of cometary gas and dust production has been completely recalibrated using ultraviolet observations from 14 comets and new, uniform visual photometry from 8 comets. The numerical work in the calibrations is presented in accompanying tables. The new theory is applied to P/Halley, using a new light curve without the artifact caused by the close approach to the earth in 1910. Gas production rates are predicted for the 1985-86 apparition. Presented at the NASA/ESA Joint Work Group Meeting at Heidelberg, 26-27 August 1981. Available as ESA Special Publication no. 174.

1145. Reinhard, R. "The Giotto Project: A Fast Flyby of

Halley's Comet." **ESA Journal,** 5 (1981), 273-85.

The GIOTTO mission is a fast flyby in March 1986 some four weeks after the comet's perihelion passage. The ten experiments to be carried out are described--they are needed to unravel the complex physical and chemical processes that occur in comet.

1146. Sekanina, Zdenek. "Properties of Dust Particles in Comet Halley from Observations Made in 1910 During Its Encounter with the Earth." in **The Comet Halley Dust and Gas Environment.** Paris: European Space Agency, 1981. pp. 55-65.

Analysis of observations of the dust tail of Halley's Comet about the time of the earth's transit across the orbit plane of the comet on 19 May 1910 yields information on particle properties which has so far been inaccessible by other means. The earth grazed the tail, whose apparent length reached up to 150 degrees. The comet's acceleration indicates the presence of submicron-sized, strongly absorbing particles. Apparently because of a low reflectivity and/or a low spatial density, their detectable contribution to the tail's brightness was limited to the immediate proximity if the orbit plane.

1147. Stanton, R.H. "CCD Tracker for Closed-Loop Instrument Pointing at Halley's Comet." New York: American Institute of Aeronautics and Astronautics. 1981.

This paper was given at AIAA 19th Aerospace Sciences Meeting, St. Louis, Missouri, 12-15 January 1981. The paper may be ordered from the AIAA as Technical Paper no. AIAA-81-0088.

1148. Trogus, Wolfgang et al. "A European Probe to Comet Halley." **Advances in Space Research,** 1 (1981), 131-36.

A European space probe to Halley's Comet is proposed. The model payload consists of 8 scientific instruments. The flyby of the Comet Halley nucleus will take place on 28 November 1985 at a about 500 km distance. The main spacecraft serves as a relay link to transmit observed data to the earth. For the probe itself, a modified ISSE 2 design is proposed.

1149. Wekhof, A. "Negative Ions in Comets." **Moon and the Planets,** 24 (1981), 157-73.

The primary source of negative ion sources in comets is discussed. Negative ion abundance for Halley's Comet has been estimated to be from $10/-6$ to $10/-10$ of electron densities. However, this ratio may be due more to formation of clusters of $A-(H_2O)n$.

1150. Whipple, F.L. "Note on the P/Halley Nucleus." in **Scientific and Experimental Aspects of the Giotto Mission.** Paris: European Space Agency, 1981. pp. 101-03.

1151. Lenorovitz, Jeffrey M. "ESA Rejects NASA Bid on Joint Halley Mission." **Aviation Week and Space Technology,** 114 (12 January 1981), 20.

   NASA proposed that it assume principal investigator responsibilities for two instruments on ESA's GIOTTO mission in exchange for ESA use of NASA's deep space network and other U.S. supplied mission services. ESA said no. Some view this rebuff as a blow to U.S. prestige.

1152. Yeomans, Donald K. **The Comet Halley Handbook: An Observer's Guide.** Jet Propulsion Lab: Pasadena, CA., 15 January 1981. 49 pp.

   The orbit of Comet Halley is described as well as expected physical behavior (brightness, tail lengths, coma diameters) in 1985-86. Provides a complete ephemeris for 1982-87, charts of comet's location from different latitudes of observation. See entry 1275 for the second edition. Available from NTIS, N81-16964/1. NASA report no. NASA-CR-163876. Jet Propulsion Laboratory publication 400-91. SUDOC classification is NAS1.12/7:400-91.

1153. Washburn, Mark. "In Pursuit of Halley's Comet." **Sky and Telescope,** 61 (February 1981), 111-13.

   A short discussion about the nature of comets. Explores the reasons for sending an intercept mission to Halley's Comet. Color drawings of spacecraft for Halley mission are provided as well as an imaginary view of a cometary nucleus close to the sun. There is also a diagram of the relative trajectories of the spacecraft, sun, and earth.

1154. "European Payload." **Aviation Week and Space Technology,** 114 (2 February 1981), 23.

   Announces that the European Space Agency has selected an all-European science payload management team for its GIOTTO Halley's Comet intercept mission thereby confirming the agency's decision not to pursue a joint cooperative effort with the U.S.

1155. Lenorovitz, Jeffrey M. "ESA's Halley's Comet Payload Studied." **Aviation Week and Space Technology,** 114 (16 February 1981), 101-02.

   Discusses the instrumentation and planned experiments aboard ESA's Halley mission. Although NASA will not have a full cooperative share in the GIOTTO project, the U.S. will be involved through coinvestigators in some 30 projects.

1156. "A Last Chance for Halley." **Astronomy,** 9 (March 1981), 59-60.

Discusses NASA Technical Report no. 82386 (see entry 1129) which emphasizes that Halley's Comet provides "our only opportunity to carry out a direct investigation of a bright active comet until Halley's next return in 2061. There is a drawing of proposed Halley intercept spacecraft.

1157. Dooling, Dave. "Europe's Comet Mission Proceeds Despite U.S." **Space World,** R-3-207 (March 1981), 13-14.

Reprinted from the **Huntsville, Alabama Times.** The ESA mission is planned for launch in July 1985. Eight months later it would pass through the comet's tail 600 miles behind the comet's head at a speed of 42 miles per second. Discusses NASA sharing in project.

1158. Marsden, Brian G. "Comet Halley and History." **1982 NASA Authorization, Vol. V. House Committee on Science and Technology serial No. 97-8.** Washington, D.C.: U.S. Government Printing Office, March 1981. pp. 3226-42.

The author is from the Harvard-Smithsonian Center for Astrophysics. Offered in support of a United States mission to Halley's Comet, Marsden presents an overview of the comet's astronomical and cultural history. Includes illustrations. SUDOC classification Y4.Sci2:97/8. See also entry 1076.

1159. National Aeronautics and Space Administration. "Comet Science Working Group Report On the Halley Intercept Mission." **1982 NASA Authorization, Vol. V. House Committee on Science and Technology serial 97-8.** Washington, D.C.: U.S. Government Printing Office, March 1981. pp. 3255-78.

Testimony offered in support of a United State mission to Halley's Comet. The proposed Halley's Comet flyby would take place at over 130,000 miles per hour and produce over 8000 images of the comet, for the first time giving us pictures of a cometary nucleus. Addresses the uniqueness of Halley's Comet, the value of a mission to the comet, the mission concept, the mission itself, imaging, in-situ experiments, non-imaging remote sensing, and post-mission options.. NASA Technical Memorandum of November 1980. Includes illustration. SUDOC classification Y4.Sci2:97/8.

1160. "Halley Mission Discussion Planned." **Aviation Week and Space Technology,** 114 (2 March 1981), 22.

Planning by European Space Agency, USSR, and Japan for
their Halley missions is described.

1161. Hughes, David W. "The Return of Halley." **Nature,**
290 (26 March 1981), 290-91.

The comet was last seen at end of May 1911.   Its
aphelion was reached in 1948, since that time the comet
has been on its way back toward the earth. The search
for the comet began in the U.S. in November 1977 with
no luck to date.   The predicted date of perihelion
passage is 9 February 1986.

1162. "Chasing Halley." **Science News,** 119 (11 April 1981),
228.

About the recent announcement of a privately organized
"Halley Fund" to gather contributions and support for a
privately financed U.S.   mission to the sky's best
known comet.   Without a U.S. mission, the coming of
the comet may mark a significant upgrading of Soviet
capabilities in planetary research.

1163. Redfearn, Judy. "Kamikaze to Halley." **Nature,** 290
(23 April 1981), 621.

The Japanese Space Agency will cooperate with the ESA
and the Russians on missions to Halley's Comet. The
Japanese craft will be the smallest of the three (135
kg.) and will photograph in the hydrogen Lyman alpha
line.

1164. "French Participation in Soviet Venus-Halley Mission
Modified." **Aviation Week and Space Technology,** 114 (27
April 1981), 152.

The Soviet desire to expand a Venus mission to include
a Halley's Comet flyby has affected plans to release
French balloons into the Venusian atmosphere.   The
Soviets have given the French assurance that their
participation will continue to be important to the
Venus/Halley mission.

1165. "Europeans, Soviets Plan Halley Data Exchange."
Aviation Week and Space Technology, 114 (27 April 1981),
148.

The European Space Agency and the Soviet Union are
working toward a data exchange agreement that could
enable ESA to make final targeting updates for its 1986
GIOTTO comet encounter utilizing information gathered
by the Soviet's comet spacecraft.

1166. Calder, Nigel. "The Comet Is Coming!" **Science
Digest,** 89 (May 1981), 23-28.

Adapted from his book of the same title (see entry
1139). Presents general comet lore, dwelling on that
associated with Halley's Comet. "Comets drive people
dotty ... in the 1980s there is solemn assertion that
Halley causes influenza." Halley will be closest to the
earth on 11 April 1986. The best view from the
Northern Hemisphere, however, may be during the last
few days of April 1986. "Halley is Halley, item number
3 in the Solar System, after the moon and the rings of
Saturn." Includes photograph and charts.

1167. European Space Agency. **The Comet Halley Probe.
Plasma Environment.** Washington, D.C.: National
Aeronautics and Space Administration, May 1981. 105 pp.

Proceedings of a Workshop Series held at Heidelberg and
Noordwijk, Netherlands, 1979-1980. Report no.
ESA-SP-155. Available from NTIS, N81-34113/3.

1168. O'Toole, Thomas. "... Not to Mention Halley's
Comet." **Washington Post Magazine,** 24 May 1981, 8-14.

A popularized gloss of Halley's Comet lore.

1169. "Looking for Halley's Comet." **Sky and Telescope,** 61
(June 1981), 500-01.

In essence, this is a review of Yeomans' **Comet Halley
Handbook** see entry 1152). Reprints three diagrams from
the **Handbook** giving the comet's positions in a dark sky
at various latitudes of observation.

1170. Tatum, Jeremy B. "The Altitude of Comet Halley."
Observatory, 101 (June 1981), 84-85.

Letter. Points out that in order to best know from
where to observe the next apparition, one must know
whether the comet is north or south of the sun rather
than whether it is north or south of the equator.
"Hasty thought might lead to the conclusion that when
the comet is south of the equator it is best to travel
to the Southern Hemisphere to see it. This could be
disasterous if the sun were even farther south than the
comet." A chart is presented that gives altitudes at
sunrise and sunset. Although the comet is south of the
equator after 23 December 1985, the Northern Hemisphere
is nevertheless the best place from which to see the
comet until about the middle of February
1986--especially the two weeks before perihelion
passage when the sun is farther south than the comet.

1171. Ferris, Timothy. "Look Out Below!" **New York,** 14 (1
June 1981), 50.

Review of Nigel Calder's, **The Comet Is Coming! The
Feverish Legacy of Mr. Halley,** 1981 (see entry 1139).

A favorable review of Calder's book. "We may laugh at Calder's collection of comet nonsense, as at the claim that comet's herald doom. But genuine doom could result should a comet blunder into the earth. ... There is no chance that Halley will hit us next time, but we will not be able to rule out the danger until we have improved our surveillance of the solar system far beyond what is permitted by, say, the Reagan budget."

1172. "Northerner's Will See Halley's Comet." **New Scientist,** 90 (11 June 1981), 685.

Relates from J.B. Tatum's letter in **Observatory** (see entry 1170) which concluded that the comet will be north of the sun for two weeks before perihelion, therefore it will be best seen then from latitudes between 30 and 70 degrees NORTH during the last week of January and first week of February 1986.

1173. Rich, Vera. "All Aboard for Halley." **Nature,** 291 (18 June 1981), 528.

French participation in the planned Soviet-French Halley probe will be largely reduced to the first stage of the mission, the flyby of Venus.

1174. Terry, Sara. "Will U.S. Miss Halley?" **Christian Science Monitor,** 24 June 1981, 1.

Column 1. NASA must find 25 million dollars in the next few months if it wants to loft even a scaled-down Halley's Comet mission in 1985.

1175. Kane, Van R. "Come Explore a Comet." **Astronomy,** 9 (July 1981), 24-25.

Editorial. Nowhere have short-sighted funding cuts had a clearer negative effect than with the proposed mission to explore Halley's Comet. Adding the mission would have increased NASA's funding by only 1 percent. Suggests supporters donate to a Halley's Comet fund to show support for the mission.

1176. McLaughlin, William I. "The Natural History of Halley's Comet." **Journal of the British Interplanetary Society,** 34 (July 1981), 267-80.

Discusses the objectives of space probe exploration. Includes tables presenting historical, physical, and orbital data for the comet and compares the different intercept missions; diagrams the paths of the intercept spacecraft. Halley's Comet must be a young comet to be still observable after so many apparitions.

1177. Wood, Lincoln J. "Navigation Accuracy Analysis for An Ion Drive Flyby of Comet Halley." **Journal of Guidance** and **Control**, 4 (July 1981), 369-75.

Presents navigation accuracy analysis results for the Halley/Tempel 2 phase of the proposed NASA space probe. Orbit determination and guidance accuracies are given for the baseline navigation strategy along with results of sensitivity studies.

1178.   Flock,   W.L.   **Telecommunications   in   Cometary Environments.**   Pasadena,   CA: Jet Propulsion Lab, 1 July 1981.   43 pp.

Propagation effects on telecommunications in cometary environment include those due to dust, inhomogeneous plasma of the coma and tail, and ionization generated by impact a neutral molecules and dust on the spacecraft. Available from NTIS, N82-11340/8.   Report no.   NASA-CR-164956.

1179.   Dickson, David.   "Halley Again?" **Nature**, 292 (9 July 1981), 102.

There is a faint glimmer of hope that the U.S.   Halley project may still be alive.   Reviews congressional action/inaction on bills funding cometary research.

1180.   Wilford, John Noble.   "Rousing U.S.   Science to Meet Halley's Red Dawn." **New York Times**, 19 July 1981, IV, 22.

Columns 3-5.   NASA has no current plans for a flight to examine   Halley's   Comet   because   of   the   Reagan administration's   budget   cuts.   The   priority   of developing   the Space Shuttle has left little money for other space projects.

1181.   "In Short." **Christian Science Monitor**, 27 July 1981, 24.

Today's question is whether   the   U.S.   will   use   its space   technology   to   make   the   kind   of   close-in observation of this spectacular event (Halley's   Comet) that was not possible before.

1182.   "Wanna Buy a Comet Picture?" **New York Times**, 31 July 1981, 22.

Editorial.   Is 300 million dollars   too   costly   for   a comet   mission?   Scientists   argue   that   it   is   a once-in-a-lifetime opportunity that may provide insight into   the   origins of the solar system.   However, it is hard to see that an American default is an   irreparable loss.

1183.   Hechler,   M.   and   F.W.   Hechler.   "Midcourse Navigation   for   the   European   Comet   Halley   Mission." in **Spacecraft Flight Dynamics**.   Paris: European Space Agency,

August 1981. pp. 189-95.

This paper was given at the International Symposium on Spacecraft Flight Dynamics, Darmstadt, Germany, 18-22 May 1981.

1184. McLaughlin, William I.   "Halley's Comet--Messenger From Space." **Spaceflight**, 23 (August 1981), 212-13.

Provides a basic list of Halley data. There is a brief summary of the history of observation of Halley's Comet. Also described is the preparation for rendezvous with the comet with diagrams of the Japanese and Soviet spacecraft. An erratum is provided in the October 1981 issue, p. 256.

1185. Will, George. "Meet Halley's Comet." **Newsweek,** 98 (3 August 1981), 80.

After recounting some comet lore, Will states that the Office of Management and Budget should not have killed NASA's plan to send up an intercept mission. It is not necessary to know the origin of the universe, but we should want to know.

1186. Large, Arlen J. "The Comet Halley Handbook." **Wall Street Journal,** 7 August 1981, 18.

Column 4. Reviews Calder's **The Comet Is Coming!** (see entry 1139) and Yeomans' **The Comet Halley Handbook** (see entry 1152). Both agree on the comet's meager appearance during the 1986 apparition. Yeomans says how meager, and Calder gives lore for this "orbiting trash ... sky pollution."

1187. Kronk, Gary. "Mr. Halley's Hairy Star." **Astronomy,** 9 (September 1981), 16-22.

Presents an astronomical history of the observation of the comet. Gives a predicted timetable for observation during the coming apparition. Quotes Brian Marsden as predicting a tail of 20-40 degrees. The comet will be far south therefore a tail of 5 degrees on the southern horizon in the Northern Hemisphere will be all that will be visible.

1188. McNiece, Mili Ve. "Halley's Comet." **North American Review,** 266 (September 1981), 25.

Poem. Prose poem describing reaction to the comet's 1910 apparition.

1189. Singer, S. Fred.   "Halley's Comet Should Shower Earth with Sub-micron Particles." **Astronautics and Aeronautics,** 19 (September 1981), 60-61.

8. The logo of the International Halley Watch, an organization to coordinate professional and amateur observation over a broad range of studies during the 1985/86 apparition. Reproduced from the frontispiece of entry 1191.

It has been discovered that during certain meteor showers (those associated with comets) there are a greater number of submicron particles. Data is sparse. Visualizes an experimental program to observe the effect of Halley's Comet on such particles.

1190. Bergstralh, Jay T. "Near-Perihelion Observations of Comet Halley from Shuttle Orbiter." in **Modern Observational Techniques for Comets.** Washington, D.C.: National Aeronautics and Space Administration, 1 October 1981. pp. 240-48.

Planned intercept missions to Halley's Comet will leave important gaps in the observational coverage of the comet's activity, especially around the time of perihelion passage. A cometary instrument package of modest size could be assembled to share space in the Space Shuttle's cargo bay with other payloads; this approach should be economical enough to permit scheduling such a package for several flights during Halley's apparition. This paper was presented at a workshop held at Goddard Space Flight Center, Greenbelt, Maryland on 22-24 October 1980. Jet Propulsion Laboratory publication 81-68. SUDOC classification is NAS1.12/7:81-68.

1191. Friedman, L. and R.L. Newburn. "The International Halley Watch: A Program of Coordination, Cooperation and Advocacy." in **Modern Observational Techniques for Comets.** Washington, D.C.: National Aeronautics and Space Administration, 1 October 1981. pp. 313-14.

"This coupling of public enthusiasm and the scientific importance is a rare commodity that we must use as an opportunity for space science when we have our once per lifetime chance in 1985-86." The IHW was conceived by (Friedman) as a small, core organization dedicated to advocating, assisting, coordinating, and ultimately achieving a large worldwide effort to study Halley's Comet by every means possible and to help present these activities to an interested public. This paper was presented at a workshop held at Goddard Space Flight Center, Greenbelt, Maryland on 22-24 October 1980. Jet Propulsion Laboratory publication 81-68. SUDOC classification is NAS1.12/7:81-68.

1192. Klinglesmith, D.A., III. "The Interactive Astronomical Data Analysis Facility--Image Enhancement Techniques Applied to Comet Halley." in **Modern Observational Techniques for Comets.** Washington, D.C.: National Aeronautics and Space Administration, 1 October 1981. pp. 223-31.

Describes a computer system designed to permit the scientist easy access to data in both visual and graphic representations. There are seven color plates included to illustrate image enhancement. This paper was presented at a workshop held a Goddard Space Flight Center, Greenbelt, Maryland on 22-24 October 1980. Jet Propulsion Laboratory publication 81-68. SUDOC classification NAS1.12/7:81-68.

1193. Rahe, Jurgen. "Plans for Comet Halley." in **Modern Observational Techniques for Comets.** Washington, D.C.: National Aeronautics and Space Administration, 1 October 1981. pp. 277-83.

Describes the various Halley's intercept missions comparing the spacecraft, communications, experiments, and other pertinent characteristics. This paper was presented at a workshop held at Goddard Space Flight Center, Greenbelt, Maryland on 22-24 October 1980. Jet Propulsion Laboratory publication 81-68. SUDOC classification is NAS1.12/7:81-68.

1194. Reinhard, R. "The ESA Mission to Comet Halley." in **Modern Observational Techniques for Comets.** Washington, D.C.: National Aeronautics and Space Administration, 1 October 1981. pp. 284-312.

The GIOTTO mission is the "remnant" of a much more ambitious joint mission with NASA involving a fast flyby in 1985. The mission could not be carried out because the necessary low thrust propulsion system was not funded by NASA. The GIOTTO mission's scientific objectives are given, this followed by a somewhat more detailed description of the 10 GIOTTO scientific instruments. The principles are explained on which the experiments are based, and the experiment key performance data are summarized.

1195. Wood, H. John and R. Albrecht. "Outburst and Nuclear Breakup of Comet Halley." in **Modern Observational Techniques for Comets.** Washington, D.C.: National Aeronautics and Space Administration, 1 October 1981. pp. 216-29.

Computer processing of five plates of Halley's Comet taken during the 1910 apparition shows that on 24 May 1910 strong asymmetric (with respect to the tail axis) fountain-like parabolic plumes had developed on the sunward side of the nucleus. Visual observations showed that after an initial fading while passing in front of the sun, the brightness increased to about magnitude 1. On the plates taken on 31 May 1910 the nucleus is clearly divided into at least three parts of nearly equal brightness. However, the last plate, taken on 3 June 1910, shows a symmetrical coma with a small stellar-like nucleus. This paper was presented at a workshop held at the Goddard space Flight Center, Greenbelt, Maryland on 22-24 October 1980. Jet Propulsion Laboratory publication 81-68. SUDOC classification is NAS1.12/7:81-68.

1196. Waldrop, M. Mitchell. "Down to the Wire on Halley." **Science**, 214 (2 October 1981), 35.

Given the mood of fiscal austerity in Washington these days, the odds of the Reagan administration's financing a 300 million dollar spacecraft are slimmer than ever. Proponents stress national prestige and scientific benefit in a last ditch effort for approval.

1197. Lenorovitz, Jeffrey M. "Soviets Express Interest in Joint Comet Program." **Aviation Week and Space Technology,** 115 (5 October 1981), 65, 69.

The Soviet Union is supportive of a cooperative scientific study of Halley's Comet. ESA's Halley intercept mission is diagrammed. Most interesting elements of the data will be obtained just before and after the spacecraft's closest approach to comet nucleus--if the craft survives.

1198. Maran, Stephen P. "Getting Ready for Halley." **Natural History**, 90 (November 1981), 32-39.

Scientists have been looking for Halley's Comet for the last three years. Discusses various plans for space exploration of the comet. The comet's nucleus shrinks with each visit and one day will no longer exist--it grows fainter with each visit. During the coming apparition it will be brightest in February 1986.

1199. Sekanina, Z. "Distribution and Activity of Discrete Emission Areas of the Nucleus of Periodic Comet Swift-Tuttle." **Astronomical Journal**, 86 (November 1981), 1741-73.

Effects of the spin rate and gravity field of a comet on the mode of outgassing are discussed and the prospects for application of the developed model to the periodic Comet Halley and other comets are briefly discussed.

1200. Yeomans, Donald K. and Tao Kiang. "The Long-term Motion of Comet Halley." **Monthly Notices of the Royal Astronomical Society**, 197 (November 1981), 633-46.

Reviews the history of efforts to determine past appearances of the comet from historical records and provides computed perihelion passage dates and orbital elements back to 1404 B.C.

1201. Overbye, Dennis. "Great Balls of Fire." **Discover**, 2 (December 1981), 20-24 ff.

A light review of scientists' anticipation of the

return of Halley's comet in 1985-86.

1202. Pournelle, Jerry. "The Halley Dilemma." **Analog**, 101 (7 December 1981), 81-84.

Poses the question as to whether a Halley's Comet space probe by the U.S. is really worth it (300 million dollars). Concludes that, on balance, it is worth doing—relatively; it doesn't cost much, it will keep imaging teams together, it serves as practice for future missions, and the U.S. looks good.

1203. Salisbury, David F. "Halley's Comet May Still Be Probed by U.S., Despite NASA Budget Cuts." **Christian Science Monitor**, 21 December 1981, 6.

Column 1. Discusses the possibility of diverting the International Sun-Earth Explorer (ISSE) satellite for a Halley flyby. Some solar scientists object. ISEE could not send back beautiful pictures, but it could make observations.

1204. "Waiting for Halley's Comet." **New York Times**, 22 December 1981, C 2.

Column 3. Dr. Fred L. Whipple says that on 2 December electronically intensified observing with 158-inch reflectors at Kitt Peak failed to detect the comet as have earlier efforts begun in 1977. If the comet is not seen in next few weeks, it may not be seen for another year.

1982

1205. Belton, Michael J.S. "The Hairy Stars." in **Science Year**. Chicago: World Book, 1982. pp. 58-69.

Much of the article deals with the search for and plans to observe Halley's Comet during its forthcoming apparition.

1206. Buffoni, L. et al. "Halley's Comet: Energy and Perturbations." **Astronomy and Astrophysics**, 108, No. 1 (1982), 141-42.

Mechanical energy and the planetary perturbations (always very weak) of Halley's Comet are provided. In the 1986 passage there will be a slight decrease in the mechanical energy of the comet with a corresponding decrease of the period.

1207. Burton, W.M. **Cometary Particle Impact Simulation Using Pulsed Lasers**. Chilton, England: Science and Engineering Research Council, 1982. 24 pp.

Calibration of dust impact detection system experiment momentus sensors for GIOTTO mission to Halley's Comet requires lab simulation of impacts at 68 km. sl for particle mass values .000001 to .0000000001 g. Available from NTIS, PB82-226507. Report no. RL-82-030.

1208. Coradini, M. et al. "The Effects of Microparticle Hypervelocity Impacts on Polished Surfaces: Tests for the Choice of the Halley Multicolor Camera." in **NATO Advanced Study Institute Series (C)**, 85 (1982), 367-87.

Discusses the testing of the multicolor camera to be used for the Halley Comet probe via quartz microparticle hypervelocity impacts.

1209. Encrenaz, T., et al. "A Theoretical Study of Comet Halley's Spectrum in the Infared Range." **Icarus**, 51 (1982), 660-64.

Synthetic spectra of Halley's Comet are calculated on the basis of current cometary models. This study shows which molecules will most likely be detected in the infrared range--including water, carbon monoxide, carbon dioxide, amonia.

1210. Felenbok, P. et al. "Deepsounding with Electrographic Camera at the Prime Focus of the CFHT--Upper Limit to the Visual Brightness of Comet P/Halley During 1981-1982 Opposition." **Astronomy and Astrophysics**, 113, No. 1 (1982), L1-L2.

Reports on the attempt to recover Halley's Comet during its 1981/1982 opposition using an electronographic camera. The unsuccessful search leads to significant constraints on the albedo and radius of the comet nucleus.

1211. Ferrin, I. "On the Brightness of Halley's Comet." **Astronomy and Astrophysics**, 107, No. 1 (1982), L7-L8.

Halley's Comet was not a "well behaved" comet photometrically in its 1910 apparition. Its light curve shows fluctuations of long and short duration that suggest that this comet has a surface with uneven distribution of volatiles.

1212. Formisano, V. et al. "The Role of the Critical Ionization Velocity Phenomena in the Production of Inner Coma Cometary Plasma." **Planetary and Space Science**, 30 (1982), 491-97.

Discusses the mechanism of ionization in the gases in Halley's Comet.

1213. Glasmachers, A. et al. "Electronic System for an

Impact Mass Spectrometer with High Dynamic Range for the GIOTTO Probe." **IEEE Transactions on Nuclear Science**, NS29 (1982), 160-63.

> The measurement objectives of the space probe GIOTTO include the analysis of the dust particles of Halley's Comet with an impact mass spectrometer of the "time-of-flight" type. Discusses performance, weight and power specifications.

1214. Newburn, R.L. and D.K. Yeomans. "Halley's Comet." **Annual Review of Earth and Planetary Sciences**, 10 (1982), 297-326.

> Covers the one truly quantitative aspect possible for Halley's Comet, its motion, in a step-by-step historical development. There is a brief discussion of the meteoroid streams associated with the comet. Compares the 1910 observation of the comet with plans for observation of the 1985-86 apparition.

1215. Reinhard, R. "Space Missions to Halley's Comet and Related Activities." **ESA Bulletin**, No. 29 (1982), 68-83.

> Describes the various comet missions (ESA, two Russian, and Japanese) in terms of spacecraft, flyby date, flyby velocity, minimum distance in the orbital plane from the cometary nucleus, phase angle to sun at which coma will be traversed, and the distance above/below the orbital plane at the closest approach to the comet. The scientific paylod is discussed accompanied by a tabular rendition of experiments to be carried out by each mission.

1216. Trendelenburg, E.A. "The Need for Inter-Agency Collaboration on Missions to Halley's Comet." **ESA Bulletin**, No. 29 (1982), 66-67.

> With four space agencies planning missions to Halley's Comet there is a need for inter-agency communication, and if possible, cooperation. Outlines problems that will be encountered on the Halley mission that have never before been encountered. NASA will be contributing significantly to the success of the operation by funding several ground-based observational networks on the international level, which will provide the necessary link between earth-based and in-situ observations and the comet ephemeris.

1217. **Halley's Comet Watch '86.** 1 Smith Court, Box 188-T, Vincentown, New Jersey, 08088. May-June 1982- .

> Newsletter. The "official publication of Halley's Comet Watch '86," described as "an international organization dedicated to historical, scientific, and sociological study of the 1985-86 return" (more

historical and sociological than scientific), was founded by Joseph Laufer. Of particular interest is the regular feature of letters from observers of the 1910 apparition. Also available from Halley's Comet Watch '86 are T-shirts, books on Halley's Comet, commemorative coins, tokens, and buttons, teaching materials and models, and bumper stickers. See also entry 1290.

1218. **The International Halley Watch.** Pasadena, California: Jet Propulsion Lab, Caltech, 1982- .

Newsletter. Comes out irregularly or quarterly. Issue No. 1 was 1 August 1982. The newsletter is the publication of the International Halley Watch, a network of scientists interested in observing the 1985-86 apparition of the comet. Serious amateurs can obtain information on IHW from S.J. Edberg, MS T-1116, Jet Propulsion Lab, 4800 Oak Grove Dr., Pasadena, CA 91109 (as cited by Freitag, see entry 1263).

1219. Asimov, Isaac. "The Long Eclipse." **Fantasy and Science Fiction,** 62 (January 1982), 137-47.

Gives historical events associated with Halley's Comet apparition by apparition. Suggests that the appearance of the comet in 11 B.C. may have been the "Christmas star over Bethlehem." Discusses characteristics of the comet, all against the backdrop of the current interest in Comet Kohoutek.

1220. Ney, Edward P. "Visibility of Comet Nuclei." **Science,** 215 (22 January 1982), 397-98.

Photographs of the Comet Halley nucleus is the goal of several planned space missions. The nucleus surrounded by a cloud of dust that is optically thick and will prevent observation of the nuclear surface, however, it can be imaged by broadband photometry.

1221. "ESA Develops Halley's Comet Spacecraft." **Aviation Week and Space Technology,** 116 (15 February 1982), 133.

Presents a detailed drawing of the general configuration of ESA's GIOTTO spacecraft. The spacecraft is equipped with a two-piece bumper shield to extend the spacecraft's life as it approaches Halley's Comet.

1222. "Search for Halley's Comet Stepped Up at Kitt Peak." **Astronomy,** 10 (March 1980), 69.

Believing that Halley's Comet can now be picked up by our largest telescopes, astronomer Michael Belton is stepping up his five-year effort to rediscover Halley's Comet. Belton expresses his disappointment that U.S.

will not mount a comet mission.

1223.   Winter,   Patty.   "Franco-Soviet   Team   Plans
Venus/Halley Mission." **Astronomy,** 10 (March 1982), 67.

Each Halley encounter probe will be three-axis
stabilized with a scan platform that will allow
flexibility in pointing. This is necessary because
scientists are uncertain of the comet's exact orbit.
Also discusses the strategy and politics of the
mission.

1224. "Two Spacecraft Venus, Halley Mission Planned."
Aviation Week and Space Technology, 116 (8 March 1982),
278-79.

The Soviet Union will use two similar spacecraft that
will deploy a probe into Venusian atmosphere and then
continue on for flybys of Halley's Comet in March 1986.
Provides diagrams of the Soviet Venera-Halley
spacecraft and rendezvous with comet.

1225. Covault, Craig. "NASA to Study Diverting Satellite
to Meet Comet." **Aviation Week and Space Technology,** 116 (22
March 1982), 22-24.

NASA considers using the International Sun-Earth
Explorer (ISEE 3) for a rerouting that would send it by
Halley's Comet. The satellite carries no imaging
equipment but would be able to characterize the comet
from a particles and fields standpoint.

1226. Davies, Owen. "SPACE." **Omni,** 4 (April 1982), 20 ff.

About an idea to send a private mission to probe
Halley's Comet. The scientific part of the mission
would catalog asteroids, while the filming of the comet
would be sold to movie houses and advertisers to
produce a profit in addition to covering the mission's
cost.

1227. Valentry, Duane. "The Worst of Halley's Comet."
Modern Maturity, 25 (April 1982), 69-71.

"A cosmic chronology of the 'troubles' it has caused"
with a prediction that some will cower in doomsday fear
and others will plan their futures around it, the
author catalogs catastrophes superstitiously associated
with comet. This article itself cultivates fear.

1228. Bobowski, Rita. "Living History." **Space World,**
S-5-221 (May 1982), 25-26.

Presents an anecdote of 1910 appearance from an
observer of the comet. From the oral history project
done in the 1970s by the American Institute of Physics

in New York City (project was funded by National Science Foundation and Sloan Foundation).

1229. Wood, Lincoln J. "Navigation Accuracy Analysis for a Halley Intercept Mission." **Journal of Guidance and Control,** 5 (May 1982), 300-06.

Describes the navigation system and navigation strategy that would have been employed on a Halley intercept mission assuming launch in the summer of 1985 and an arrival in March 1986. Spacecraft and comet relative orbit determination accuracies given.

1230. Waldrop, M. Mitchell. "In Quest of Comet Halley." **Science,** 216 (7 May 1982), 606.

Somewhere between the orbits of Uranus and Saturn, still more than a billion kilometers out, a mountain-sized fragment of ice and rock is falling toward the sun; it is about magnitude 25 or 26 in brightness. Comet nuclei are unknown because they are hidden by vapors.

1231. Hendrie, Michael J. "The Return of Halley's Comet." **Spaceflight,** 24 (June 1982), 242-48.

Review of the comet's history and plans for observation in 1986. Includes illustration and portraits.

1232. Morris, Charles S. and Daniel W.E. Green. "The Light Curve of Periodic Comet Halley 1910 II." **Astronomical Journal,** 87 (June 1982), 918-23.

Photometric parameters for the comet have been derived from 144 total visual magnitude estimates. Based on these results, a forecast of the visual brightness of the comet's 1985-1987 apparition is presented--at its brightest the comet will be at 3.9 magnitude on 10 April 1986.

1233. Eberhart, J. "Bound for Halley: Surviving the Unknown." **Science News,** 121 (12 June 1982), 391.

Discusses the dangers to, and prospects for survival of, the close-flyby spacecraft that will attempt to take measurements, such as direct samplings of the comet's enveloping coma and tail. The jet pattern of the comet is complex and irregular.

1234. "Payload Incentives." **Aviation Week and Space Technology,** 116 (21 June 1982), 63.

Arianespace is offering a launch price incentive (a one million dollar reduction) to insure that it has a payload to be orbited with the European Space Agency's GIOTTO mission to Halley's Comet. The typical launch charge is approximately 25-30 million dollars.

1235. Tatum, Jeremy B. "Halley's Comet in 1986." **Mercury,** 11 (July 1982), 126-31.

Outlines our history of understanding of Halley's Comet and predicts those places on earth from which the comet will be best visible in 1986. Presents diagrams indicating the position of the comet, its distance from the sun, and best observation latitudes. The magazine's cover has a photograph of the comet taken 8 May 1910 at Mt. Wilson.

1236. "Seven Nations to Contribute to Soviet Comet Observation." **Aviation Week and Space Technology,** 117 (12 July 1982), 72.

The Soviet Union's unmanned flyby missions to Halley's Comet will carry experiment payloads developed in cooperation with France, West Germany, Austria, Czechoslovakia, Hungary, Bulgaria, and Poland. Describes what the scientific instruments are to be.

1237. Belton, M.J.S. and Harvey Butcher. "Limits on the Nucleus of Halley's Comet." **Nature,** 298 (15 July 1982), 249-51.

Reports a limiting magnitude to the brightness of the nucleus of Comet Halley. The comet may not be recovered until early November 1984, but could be found as early as September or early October 1982.

1238. "Halley's Comet--Out of Sight, But Not Beyond Study." **New Scientist,** 95 (29 July 1982), 300.

About the Belton and Butcher article (see entry 1237).

1239. Brady, J.L. "Halley's Comet: A.D. 1986 to 2647 B.C." **Journal of the British Astronomical Association,** 92 (August 1982), 209-15.

In earlier predictions of the apparitions of the comet a three or four-day error, always with the same sign has appeared consistently. Using backward integration, the author ventures to make a forward integration and to predict the 1986 return.

1240. Bloom, Britton. "Grand Preparations Afoot for the Return of Halley's Comet." **Christian Science Monitor,** 12 August 1982, 17.

Column 1. Astronomers are preparing the largest astronomical project ever organized to view a celestial visitor. Reports on the meeting of International Halley Watch at Patras, Greece and sketches out some of the preparations underway.

By Albert J. Forbes

Grand preparations afoot for the return of Halley's comet

9. Cartoon by Albert J. Forbes, 12 August, 1982. See entry 1242. Forbes in the *Christian Science Monitor* © 1982 TCSPS.

1241. Cowen, Robert C. "The Importance of NOT Seeing Halley's Comet: Research Notebook." **Christian Science Monitor,** 12 August 1982, 17.

Column 1. Astronomers Michael Belton and Harvey Butcher have made an important observation in not seeing the comet on 2-3 December 1981. Their failure has helped them to estimate the comet's size (4 km. across), mass (37.5 tons), and magnitude (no less than 24.3).

1242. Forbes, Albert J. **Christian Science Monitor,** 12 August 1982, 17.

Cartoon. Shows people on various continents peering skyward through telescopes as the comet soars overhead. Accompanies Britton Bloom article on same page (see entry 1240).

1243. Covault, Craig. "Cost, Gimbal Difficulties Threaten Halley Mission." **Aviation Week and Space Technology,** 117 (16 August 1982), 26.

Both the science mission development costs and the potential cancellation of the Advanced Gimbal System (AGS) are threatening the Halley's Comet observation schedule. Some scientists worry about over-emphasis of Halley's Comet over other objectives.

1244. "Stakeout for Halley's Comet." **New York Times,** 31 August 1982, C 4.

Column 6. An international network of scientists is being formed to observe Halley's Comet when it appears in 1986. The organization is called International Halley Watch, led by Ray Newburn of the Jet Propulsion Lab in Pasadena, California and Jurgen Rahe of the Remeis Observatory, West Germany.

1245. "Giotto Satellite Will Traverse Flight Path of Halley's Comet." **Industrial Research,** 24 (September 1982), 88-89.

Coma experiments will be made by a mass spectrometer and an ion mass spectrometer; in the particle experiments a particle impact analyzer, a dust impact detector, and Halley optical probe; plasma experiments use two plasma analyzers and magnetometer.

1246. Kane, Van R. "Bruce Murray Interview." **Astronomy,** 10 (September 1982), 24 ff.

Murray is a co-founder of the Planetary Society and past director of Caltech's Jet Propulsion Laboratory. He blames the failure of the U.S. to have a Halley's Comet mission on the weakened presidency combined with mediocre leadership in the executive branch.

1247. "NASA Decision on U.S. Halley Flyby Research Funding Nears." **Aviation Week and Space Technology,** 115 (28 September 1982, 42-43.

Discusses various alternatives for a U.S. Halley mission utilizing the Jet Propulsion Lab's Galileo Jupiter orbiter/probe system to minimize scheduling risks. The hope is to return cometary particles to earth--particles 1 mm. in size would be sufficient for study.

1248. "Group Organized to Observe Flight of Halley's Comet." **Industrial Research,** 24 (October 1982), 81-82.

Scientists gathered at the International Astronomical Union's meeting in Patras, Greece announced the formation of a global group, International Halley Watch (IHW), to coordinate observations of the comet. The organization seeks comprehensive observation from all southern stations.

1249. Dunkle, Terry. "To Catch a Comet." **Science,** 82 (October 1982), 44-52.

Describes the unsuccessful attempt in February 1982 at McDonald Observatory to sight Halley's Comet. Includes color illustration.

1249A. Rutherford, F. James, "Sputnik, Halley's Comet, and Science Education." **Forum for Liberal Education,** 5 (October 1982), [?].

The author laments the poor quality of science education in the United States. Asking, rhetorically, whether troubles in science education are merely cyclical or endemic he ponders the state of the world at the time of the last three appearances of Halley's Comet. "Not only is the world the comet returns to each trip different from the way it was the time before, but the magnitude of the difference is greater, probably exponentially so. This kind of comparison tells us two things about the world of the future. It tells us that we cannot now accurately describe the world as it will be when Halley's Comet returns in 2062. ... If human civilization is not here, a real possibility, it will be because we failed to come to terms with ourselves as the generators of knowledge and the users of technologies. ... Graduates of the next quarter century will determine whether when Halley's Comet returns in 2062 there will be a world of human beings at all, and if so what kind of world it will be." Reprinted in **Journal of College Science Teaching,** 12 (March/April 1983), 305-06.

1250. Alexander, George. "Halley's Comet Sighted: Last Seen in 1910, the Wanderer Returns." **Washington Post**, 21 October 1982, D1, 5.

Announces the first sighting by astronomers at Mt. Palomar Observatory. The comet was seen as a tiny pin point of light that moved relative to background stars. At the moment the comet is more than one billion miles away. A 1986 Space Shuttle is due to gather comet data.

1251. Wilford, John Noble. "Halley's Comet Is Sighted." **New York Times**, 21 October 1982, 16.

Column 2. After searching skies for nearly five years astronomers at Mt. Palomar on 15 October sighted Halley's Comet in the region of the constellation of Canis Major, using a 200-inch telescope with an imaging system built around a charge-couple device.

1252. "Old Reliable." **New York Times**, 24 October 1982, 20.

Column 1. Editorial. A light rendition of some historical events that have concurred with Halley's Comet--in regard to the comet's being sighted for first time since 1910. In February 1986 it will pass within 39 million miles of earth giving us all a crick in the neck. It just keeps on rolling along.

1253. Wilford, John Noble. "Halley's Comet: The Long Hello Begins." **New York Times**, 26 October 1982, C 1,3.

Column 1. The comet has been found; its regularity provides scientists an opportunity to learn more about the comet during the next five years. Gives intercept dates (8, 13 and 15 March 1986) for the three comet space probes. Presents a diagram of the comet in respect to the earth.

1254. "Halley's Comet Returns." **New Scientist**, 96 (28 October 1982), 215.

Acknowledges Donald K. Yeomans of the Jet Propulsion Lab in Pasadena, the astronomer who calculated the ephemeris used in the 1982 sighting of the comet.

1255. Eberhart, J. "Recovery of Comet Halley Reported." **Science News**, 122 (30 October 1982), 277.

Announces the resighting of Halley's Comet by David C. Jewitt and G. Edward Danielson at the Mt. Palomar Observatory. Calls the comet the most famous in existence. Includes a photograph of an object identified as the comet one billion miles away from earth.

1256. "Giotto, Halley's Comet Interceptor." **Mechanical Engineering,** 104 (November 1982), 73.

Contracts for the production of the ESA GIOTTO spacecraft have been let. GIOTTO will be Europe's first space explorer--it will take eight months to reach the comet, but the time available for observation will be only a few hours. It is intended that GIOTTO will traverse the coma region at a fly-by velocity of 68 km. per second. A camera will photograph the comet's nucleus and measurements will be made of its magnetic field.

1257. "Comet Trekking: Halley's Heavenly Body Returns." **Time,** 120 (1 November 1982), 69.

A Caltech team detected the comet on the night of 15-16 October using a charge-coupled device with the 200-inch Hale reflector telescope on Mt. Palomar. Includes illustration.

1258. "Halley's Comet Detected by Hale Telescope." **Aviation Week and Space Technology,** 117 (1 November 1982), 22.

The comet was detected by the Hale telescope at Mt. Palomar observatory by means of a series of 8-minute exposures utilizing a sensitive electronic light detector known as the prime focus universal extragalactic instrument on 16 October. The comet was about 1 billion miles away.

1259. Begley, Sharon and John Carey. "The Heavenly Streakers (Comets)." **Newsweek,** 100 (8 November 1982), 81.

Some general discussion of comets and a few historical associations of Halley's Comet (Norman conquest and fall of Jerusalem) lead into a description of the rendezvous missions planned for 1986.

1260. Dickinson, Terence. "Halley's Comet Returns to Earth." **Macleans,** 95 (8 November 1982), 60.

Reviews plans for space exploration of the comet. "The Halley challenge has been taken up by just about every nation capable of launching. Significantly absent from exploration plans is the United States"--300 million dollars funding was refused for probe.

1261. Cowen, Robert C. "U.S. Satellite Quits Sun Vigil to Intercept Two Comets." **Christian Science Monitor,** 10 November 1982, 6.

Column 1. A monitoring satellite (ISEE-3) was taken out of its orbit in September and is now on its way to meet Comet Giacobini-Zinner a half-year before it is due to flyby Comet Halley. This provides some compensation for scientists disappointed by lack of a United States Halley mission.

1262. Hughes, David W. "The Recovery of Halley's Comet." **Nature**, 300 (25 November 1982), 318.

American and European astronomers have been competing with each other to be the first to sight Halley's Comet. The American's have won, using the Hale telescope at Mt. Palomar Observatory, California. The comet is three times fainter than estimated.

1263. Freitag, Ruth S. "Halley's Comet: A Selected List of References." **Library of Congress Information Bulletin**, 41 (26 November 1982), 394-400.

A taste of the multitudinous literature devoted to Halley's Comet. There is an announcement that Freitag is at work on a comprehensive Halley's Comet bibliography.

1264. Strout, Richard L. "Halley's Comet and Other Cost-cutting Targets." **Christian Science Monitor**, 26 November 1982, 22.

Column 1. Editorial. Halley's Comet has been at the cold dark edge of the solar system all this time and it is now whizzing back for a look at us and the sun before darting off again. Its head will be 100,000 miles across and tail 50 million miles long.

1265. "Halley's Comet Swings Into View." **Discover**, 3 (December 1982), 16.

Announces David Jewitt's and G. Edward Danielson's confirmation of the return of Halley's Comet on 16 October 1982. During the past five years virtually every major observatory in the world has spent some time scanning Canis Minor for the comet.

1266. "International Halley Watch Organized." **Astronomy**, 10 (December 1982), 60.

International Halley Watch is formed and invites all comet observers, professional or amateur, to share their data with the IHW networks now being created. Lists techniques to be used by IHW experiment teams to study the comet.

1267. DiCicco, Dennis. "Comet Halley Found." **Sky and Telescope**, 64 (December 1982), 551.

Photograph of Danielson and Jewitt, first sighters of the comet during this apparition. The comet has been designated 1982i. Includes illustration and color portrait.

1268. O'Toole, Thomas. "Comet's Return Figures to Be Most Watched Event Ever." **Washington Post**, 28 December 1982, A3.

Why watch Halley's Comet? It is the only comet with large nucleus, coma, and tail to be seen this century. Also, it has swarms of exotic dust particles and complex molecules that have been part of comet since the dawn of time.

1983

1269. Branley, Franklyn M. **Comet 1986.** New York: Lodestar Books, 1983. 96 pp.

A basic non-technical description of the comet and its orbit with disproportionate attention paid to superstition and comet lore associated with Halley's Comet. Geared to an audience aged 7-to-12-years old.

1270. Cook, Anthony. "Comet Halley Update." **Griffith Observer**, 47 (February 1983), 8-9.

Notes that the recovery on 16 October 1982 by Caltech astronomers Dr. G. Edward Danielson and graduate student David C. Jewitt heralds the 29th recorded appearance of Halley's Comet since 240 B.C. The comet was found in the constellation Canis Minor close to the location predicted by Donald K. Yeomans in 1977. The recent observation of the comet is thought to be the first view ever obtained of a completely frozen nucleus. The comet was 1.015 billion miles from earth and moving along its orbit at 6.5 miles per second. At the time of perihelion on 9 February 1986 the comet's speed will have increased to 34.5 miles per second. A diagram is provided of the comet's orbit superimposed on the orbits of Neptune, Uranus, Saturn, Jupiter, Mars, and the earth (positions of 1986 perihelion, 1948 aphelion, 1982 recovery, and last 1911 sighting are marked). The front cover of this issue has a picture of the comet from the **Nuremburg Chronicle**.

1271. Edberg, Stephen J. **International Halley Watch Amateur Observers' Manual for Scientific Comet Studies.** Washington, DC: National Aeronautics and Space Administration, 1 March 1983. 2 vols.

Research performed by the CalTech's Jet Propulsion Lab under contract with NASA. Volume 1 provides instructions on the proper methods of generating meaningful scientific data on comets. Volume 2 contains an ephemeris and star charts for finding the comet. SUDOC classification is NAS1.12/7:83-16/pts. 1

246

and 2.

1272. Halliday, Ian. "Preparing for Halley's Comet." **Journal of the Royal Astronomical Society of Canada,** 77 (April 1983), 63-73.

Notes that the coming apparition will be the 29th recorded appearance of the comet. Discusses the preparations underway for observation of the comet and the timetable for their implementation.

1273. "Rendezvous with Halley's Comet Scheduled for Acoustic Sensors." **Materials Evaluation,** 41 (April 1983), 547.

Tiny sensors, which are highly sensitive microphones are part of the Dust Impact Detection System (DIDSY) project being planned by scientists from the University of Kent in England as one of the ten major scientific experiments making up mission GIOTTO. During the encounter with the comet, dust particles in the comet's tail will strike the satellite, and the DIDSY sensors will pick up the sound of that impact, sending the resulting signals back to mission GIOTTO monitoring stations on earth. Because those acoustic signatures will be related to the detected momentum of impact, they will provide potentially useful evidence about the mass of the dust particles.

1274. Yeomans, Donald K. "Comet Halley and Some Dubious Achievement Awards." **Griffith Observer,** 47 (April 1983), 2-10.

An entertaining look at errors and misjudgments that have been made and propagated in regard to Halley's Comet through the ages. Included are ten illustrations. It is mentioned that Yeomans is in the process of writing a popular book about Halley's Comet.

1275. Yeomans, Donald K. **The Comet Halley Handbook.** 2nd edition. Washington, D.C.: U.S. Government Printing Office, 15 May 1983. 44 pp.

Created for the International Halley Watch, this is an updated version of entry 1152. This edition contains an updated orbit that includes recovery observations of the comet through 14 January 1983. Improved magnitude estimates for Halley's Comet have been added, as well as a section on the comet's dust tail by Dr. Zdenek Sekanina. As with the earlier edition, information if provided on the orbit of the comet, the expected physical behavior in 1985-86, observing conditions in 1985-86 from various latitudes in both Northern and Southern Hemispheres, an ephemeris for 1982-1987, and figures showing angular elements, elliptical plane projection, relative positions of the comet and the

earth, the path of the comet on the celestial sphere
during November 1985 through may 1986, total magnitude
estimates, linear tail lengths computed from naked-eye
estimates, ground based observing conditions, predicted
brightness profiles of the dust tail on 13 March and 10
April 1986. This gem of a volume is a must for any
serious observer of the comet, be they amateur or
professional. Jet Propulsion Laboratory publication
400-91. SUDOC classification is NAS1.12/7:400-91/2.

1276. Eicher, David. "Waiting for Halley Gazer's Guide."
Astronomy, 11 (June 1983), 35-38.

"Although nothing can yet be done about observing
Halley, it's a good time to start **thinking** about
observing the comet." The comet will probably
disappoint the public and be of only the 4th magnitude
in brightness at the time of its perihelion on 9
February 1986.

1277. Cravens, Gwyneth. "Toasting Halley's Comet." **New
Yorker**, 59 (27 June 1983), 76-79.

Describes a meeting of the Halley's Comet Society
(London, founded 1976) which meets annually for fun and
toasting of the comet. The society "has no object, no
purpose, no aim, no raison d'etre, and, as such, is
rather like the United Nations."

1278. Eberhart, J. "Pioneer Venus Craft to Study Halley."
Science News, 124 (9 July 1983), 21.

Although the United States is not sending up a Halley
spacecraft per se, the Pioneer Venus spacecraft already
in orbit will be in a unique position to study the
comet at its most sunward point of orbit, when the
comet's most active outpourings will begin. As Venus
circles the sun in early 1986, Halley's Comet will be
coming around in the other direction, passing about 40
million kilometers from the planet and orbiter on 4
February--just five days before the comet reaches
perihelion.

1278A. Cioffi, Mickey. "He Launches 'Watch' for the
Comet." **Grit**, 31 July 1983, 6.

Tells the story of Joe Laufer of Vincentown, New
Jersey, who has formed Halley's Comet Watch. Laufer
first became interested in the comet when doing a
family history (both his parents were born in 1910) and
again later when doing a paper on Mark Twain. Laufer
has formed his own association, Halley's Comet Watch
'86, and is publishing a newsletter, and selling comet
memorabilia. See also entries 1217 and 1290.

1279. Freitag, Ruth S. "Elliptical Designs: Halley's

248

Comet as a Medium for Advertising Messages." **Quarterly Journal of the Library of Congress,** 40 (Summer 1983), 266-77.

Some ad designs collected in the course of compiling a comprehensive bibliography on Halley's Comet are offered from the U.S., Canada, France and Germany. This is a sampling of what has so far been unearthed at the Library of Congress.

1280. DiCicco, D. "A User's Guide to Halley's Comet." **Sky and Telescope,** 66 (September 1983), 211-12.

Discusses several of the user's guides and handbooks coming on to the market as Halley's Comet approaches. Particular attention is paid to the efforts of the International Halley Watch (IWH)--which is in the process of publishing the "International Halley Watch Amateur Observers' Manual for Scientific Comet Studies" (to be published by Sky Publishing Corp.).

1281. Eichenlaub, Jesse. "Four Probes to Comet Halley." **Astronomy,** 11 (September 1983), 16-22.

After briefly discussing past superstition associated with comets, each of the four planned Halley's Comet space probes are described: GIOTTO (European Space Agency) cylindrical in shape and standing almost 3 meters high; Planet A (Japan) is cylindrical in shape and stands about .7 meters tall and 1.4 meters wide; and Vega 1 and 2 (Soviet). In each instance the observation strategy and spacecraft accoutrement are briefly described. All the probes must be concerned about balancing a desire to get as near as possible to Halley's Comet and yet not be destroyed by it before gathering and relaying information.

1282. Hughes, David W. "Temporal Variations of the Absolute Magnitude of Halley's Comet." **Monthly Notices of the Royal Astronomical Society,** 204 (September 1983), 1291-95.

Comets decay. Every time they pass the sun considerable amounts of gas and dust are emitted by the nucleus; the average cometary nucleus loses a surface layer of thickness about 100 cm. at each perihelion passage. Halley's Comet was about 0.5 +/- 0.5 magnitudes brighter 2000 years ago and is losing about 2.3 +/- 2.3 percent of its mass at each apparition.

1283. "Waiting for Halley." **Scientific American,** 249 (September 1983), 88, 92.

Notes that the comet has already been detected, but the prospects for the average spectator are not good. When the comet reaches its perihelion on 9 February 1986 the earth and the comet will be on opposite sides of the sun. In spite of this orbital inconvenience the comet is eagerly awaited by astronomers. Instrument networks and space intercept missions should yield photographs and data compensating for the comet's poor visual appearance. Counts the coming apparition as the 29th recorded appearance of the comet. Briefly summarizes the various intercept missions.

1284. "How to Switch Off a Comet." **Sky and Telescope,** 66 (October 1983), 292-93.

Other than the Japanese, Russian and ESA space missions to Halley's Comet, perhaps the most radical proposal for the study of the comet comes from Ray Norris and John Ponsonby of Manchester University's Nuffield Radio Astronomy Laboratories at Jodrell Bank in England. Their plan is to use the 250-foot Mark 1A radio telescope to transmit precisely tuned microwave signals at the comet to modify its natural radio emissions. In the process they hope to learn much about how such signals are produced and also something of the physical conditions in the comet. It should be possible to measure the magnetic field of Halley's Comet, if any exists.

1285. Tatum, J.B. and E.C. Campbell. "The Cyanogen Bands of Halley's Comet." **Journal of the Royal Astronomical Society of Canada,** 77 (October 1983), 257.

Abstract of a paper presented at the Society's annual meeting held at the University of Victoria, 26-30 June 1983. The profile of the 388.3 nm CN band, excited by the Swings fluorescence process, is calculated for Comet Halley for every half-day from 200 days before perihelion to 200 days postperihelion, using high-resolution whole disc solar irradiation spectra supplied by Kitt Peak Observatory.

1286. "Comet Coming." **London Times,** 13 October 1983, 6.

Column h. Blurb. Soviet astronomers have spotted Halley's Comet 870 million miles away, using the world's largest mirror telescope at Zelenchuk Observatory in the Caucasus.

1287. Ronan, C.A., "Halley's Comet." **Nature,** 305 (13 October 1983), 570.

Letter. Reaffirms D.W. Hughes's assertion that Halley pronounced his name as one would now pronounce "Hawley." Presents some retooled doggerel (c.f., entry 366) emphasizing this pronunciation: "Of all the comets in the sky/There's none like comet Halley,/We see it with the naked eye,/But this time rather poorly."

1288. McIntosh, B.A. and A. Hajduk. "Comet Halley Meteor Stream: A New Model." **Monthly Notices of the Royal Astronomical Society**, 205 (December 1983), 931-43.

Although the association of the Orionid and n-Aquarid meteor showers with Halley's Comet have long been recognized, a satisfactory explanation of the displacement of the meteoroid streams from the comet orbit has not been given. Utilizing the recently determined history of the comet orbit back to 1404 B.C. it is postulated that because of perturbations by the major planets rapid motion of the longitude of the nodes of the comet orbit is produced, and therefore the meteoroids simply exist in orbits where the comet was many revolutions ago. The meteor streams are bands of particles which have been ejected from the comet at each perihelion passage, their finite differential velocities causing them to lag or advance with respect to the comet position and spread around the orbit.

1289. O'Toole, Thomas. "Soviets Willing to Cooperate in Comet Tracking." **Washington Post**, 23 December 1983, 1A.

Column 3. "In a scientific breakthrough that took western countries by surprise, the Soviet Union has agreed to cooperate in all phases" of International Halley Watch. They will share all their observations of the comet, including those made by their two space probes. The Soviets have also asked the United States for help in tracking their Halley spacecraft in order that they might navigate as close as possible to the comet.

1984

1290. Laufer, Joseph. **Halley's Comet.** Pemberton, NJ: Burlington County College, 1984. 55 pp.

A booklet that was put together for a non-credit course at the college. It is an amalgam of the issues of **Halley's Comet Watch Newsletter,** May-June 1982 through February 1984 (see also entry 1217). Included are numerous letters from persons who viewed the comet in 1910. Also appended is a short bibliography compiled by the author and a complete reprint of Ruth Freitag's bibliography as it appeared in the **LC Information Bulletin** (see entry 1263). Las two pages depict t-shirts, buttons and bumper stickers that the author is selling through his Halley's Comet Watch.

1291. Bortle, John E. and Charles S. Morris. "Brighter Prospects for Halley's Comet." **Sky and Telescope,** 67 (January 1984), 9-12.

10. Advertisement for Halley's Comet Watch '86. This particular ad ran in *Sky and Telescope* in May and June 1983, and April, May, and June 1984; *USA Today* on 30 March, 1984; *The Reflector* in May 1984; *Coin World* in April 4, 11, and 18, 1984 issues. Other Laufer ads appeared in *Popular Science, Science Digest, Boy's Life, Alaska Fisherman, Country Journal, Good Old Days, Mother Earth News, Popular Mechanics, True West Magazine, Grit,* and *The New Republic*. See entries 1217, 1278A, and 1290. Reprinted with permission of Joseph Laufer.

The comet reaches its maximum intensity (brightness) two or more weeks after its closest passage to the sun--it is about 2 magnitude brighter. There is no evidence that the intrinsic brightness of the comet has lessened with time. A diagram of the positions of the comet from 1970 to 1986 is presented. The orbit and visibility of comet, what to expect in 1985-86, and prediction for the comet's brightness are all discussed. A chart depicting the number of days before and after perihelion that the comet was visible from the northern hemisphere is also included.

1292. Elson, Benjamin M. "ESA Uses Soviet Data in Comet Mission." **Aviation Week and Space Technology**, 120 (2 January 1984), 46-48.

Comet Halley ephemeris data obtained from the Soviet Venus-Halley (Vega) probes will help ESA fine-tune the GIOTTO spacecraft's close encounter with Halley's Comet on 13 March 1986. Briefly discusses each of the comet missions (Soviet, Japanese, and ESA), citing as mission objectives: imagery of the comet, with a resolution of 50 meters (164 ft.) from GIOTTO and 180 meters (590 ft.) from Vega; Identification of the composition of the gases and dust particles in the Characterization of physical processes and chemical reactions in the coma; Measurement of gas production rates and dust flux; The study of cometary plasma/solar wind interactions. GIOTTO will weigh 2,090 pounds at time of launch and 1,126 pounds at the time of encounter.

1293. "Soviets to Aid Halley's Comet Observations." **Aviation Week and Space Technology**, 120 (2 January 1984), 20.

The Soviet Union has agreed to provide data from its Vega-1 and Vega-2 space exploration probes to the International Halley Watch and other agencies studying Halley's Comet. Gives timetable for the encounters of the various missions to Halley's Comet: Vega-1 on 6 March 1986 at 6,210 miles; Vega-2 on 9 March 1986 at 1,863 miles; ESA's GIOTTO on 13 March 1986 at 62 to 310.5 miles; Japan's MS-T5 on 8 March 1986 at 621,000 miles; Japan's Planet-A on 7 March 1986 at 124,200 miles; and NASA's ASTRO-1 will be on board the Space Shuttle in March 1986 as well.

1294. "Comet Watchers." **Country Journal**, 11 (March 1984), 34-35.

"Our most famous visitor from outer planetary space will soon return." Two Soviet spacecraft, Vega 1 and Vega 2 will head for an encounter with Halley's Comet on 6 and 9 March 1986, respectively; and two Japanese spacecraft, MS-T5 and Planet-A will rendezvous with the comet on 8 March 1986. The European Space Agency's

GIOTTO craft will intercept the comet on 13 March. The U.S. spacecraft, Explorer 3, will be rerouted to observe the solar wind around Halley's tail. For earthbound observers the view will not be as impressive as it was in 1910. The tail will reach a maximum length of 50 million miles.

1295. Cook, Anthony. "Comet Halley Update No. 2." **Griffith Observer,** 48 (March 1984), 9-10.

Summarizes the observations made during the first observing season that ran from October 1982 to February 1983. The early observations of the comet indicate that the nucleus may be varying in apparent brightness. It is not yet certain whether the brightness changes are due to some property of the nucleus, such as uneven reflectivity coupled with the rotation of the nucleus or if it is due to the extreme difficulty in getting reliable measurements of such a faint object. Includes 2 negative images of the distant nucleus of Comet Halley taken with a charged coupling device camera.

1296. Kitfield, James. "The Comet Is Coming." **Omni,** 6 (March 1984), 22, 105.

After giving a little of the standard historical background, the focus is on the GIOTTO rendezvous mission which is due to blast off from the jungles of French Guiana in July 1985. Once in the air the strategy is a "onetime kamikaze"--get in as close as possible, immediately relay all data to earth and find out as much as possible before the probe dies. Calculations suggest that GIOTTO and Halley's Comet will pass each other at a velocity of 70 kilometers per second--roughly 30 times the speed of a bullet.

1297. Trachtenberg, Jeffrey A. "Spaced Out." **Forbes,** 133 (26 March 1984), 216.

About Burt Rubin, founder of Halley Optical Corp., and his "sleek 22-ounce telescope dubbed the Halleyscope." Anticipating that the telescope market will boom as it did in 1910, Rubin has already pumped $800,000 of his own into his company and is seeking private investors to put up an additional $4.5 million. The Halleyscope will cost around $230 retail. Rubin insists that "when the comet comes back, you're going to see a phenomenon, comet mania." A photograph of Rubin with the Halleyscope is included.

1298. Larson, S.M. and Z. Sekanina. "Coma Morphology and Dust-emission Pattern of Periodic Comet Halley: I. High Resolution Images Taken at Mount Wilson in 1910." **Astronomical Journal,** 89 (April 1984), 571-78.

A new image algorithm has been devised to improve

visibility of features in the head of Halley's Comet on the high-resolution photographs taken in 1910. The most striking features are spiral jets that "unwind" from the central condensation and evolve into expanding envelopes on a time scale of days. They consist of dust particles ejected continuously from discrete regions on the sunlit side of the rotating nucleus and are, in their early phase of development, essentially two-dimensional formations. The 1910 images give a preview of the type of activity that the Near-Nucleus Studies Network of the International Halley Watch will be concerned with during Comet Halley's upcoming apparition.

1299. Wilford, John Noble. "Astronomers Worldwide Rehearse for Halley's." **New York Times,** 3 April 1984, 19, 21.

The 800 astronomers from 47 countries who have agreed to participate in the International Halley Watch are taking practice observations of the Comet Crommelin. Relates the activities of the astronomers at Table Mountain Observatory, operated by the Jet Propulsion Laboratory. The comet was first spotted in October 1982 when it was three times farther away from the sun and about 2000 times fainter than it was at first sighting in 1909--"a rough measure of how far astronomy has come in seven decades."

1300. "Red Surface Seen on Halley's Comet." **New York Times,** 24 April 1984, C3.

Columns 1-2. Preliminary analysis of the first spectroscopic observation of the comet for this apparition indicates that the surface of the comet is very red in color. Astronomers suggest that this means that the frozen nucleus may be covered with a kind of primeval material that is rich in complex organic carbon-bearing molecules. These could be similar to materials found in coal tar and in certain primitive meteorites. Observations were obtained at the Kitt Peak National Observatory in Arizona.

1301. "Shower of Meteors Is Due Next Week." **St. Paul Pioneer Press** [St. Paul, Minnesota], 26 April 1984, C2.

Column 5. An Eta Aquarid meteor shower will occur next week, peaking at about midnight on 4 May, as the earth passes through debris from the tail of Halley's Comet (1910). Part of its tail is left behind on each trip through this part of the solar system.

GLOSSARY OF SELECTED ASTRONOMICAL TERMS

ANTITAIL: Projection effect, when the earth crosses the orbital plane of the comet, sometimes make a portion of the comet's tail appear to point towards the sun.

APHELION: The point in its orbit when the comet is farthest away from the sun.

APPARITION: The period of time that the comet is visible from the earth.

ASCENDING NODE: The point where the comet crosses the north side of the ecliptic.

AU (Astronomical Unit): Is the mean distance between the earth and the sun; one AU is approximately 92,960,000 miles (149, 604,970 km.). The AU is the principal unit of measurement within the solar system.

COMA: The volume containing gas and dust around the nucleus of the comet which has not yet been swept into the tails by the solar wind and solar radiation pressure.

DECLINATION: Corresponds to geographic location. It is measured in degrees from the celestial equator, being positive for stars of the Northern Hemisphere and negative for the southern stars.

DESCENDING NODE: The point at which the comet crosses the south side of the ecliptic.

DUST TAIL: Solid dust particles (from the nucleus of the comet) responding to solar radiation pressure and their orbital motion, are pushed away from the nucleus. The dust tail is seen because of sunlight scattered by the dust particles.

ECCENTRICITY: The deviation from a circular orbit.

ECLIPTIC: The intersection of the earth's orbit with the celestial sphere.

ELEMENTS: According to Cowell and Crommelin the elements for Halley's Comet in 1910 are as follows:
        a=17.94527 (the semi-major axis; half the
           longest diameter of orbit)
        e= 0.967281 (eccentricity of orbit)
        i=162 degrees 12' 42" (inclination angle where
           planes of the earth's and comet's orbits meet)
        Longitude of Node=57 degrees 16' 12"
        Node to Perihlion=111 degrees 42' 16"
        T=Time of perihelion
The first two elements give the exact size and form of the elliptical orbit; the third and fourth fix the position of the plane of the orbit, relative to the earth's orbit; the fifth element fixes the position of the ellipse in that plane; the last item is the exact time of perihelion.

EPHEMERIS (EPHEMERIDES): Tables of data giving the computed places of the comet at certain points in time, derived from the comet's orbital elements.

HEAD: The nucleus and coma of the comet are collectively referred to as the head.

HELIOCENTRIC POSITION: The comet's position in a coordinate system with the center of the sun as the point of origin.

INCLINATION: The element of orbit which indicates the angle between the plane containing the orbit of the comet and some reference plane (the plane of the earth's orbit).

ION: An ion is a neutral atom or molecule which acquires additional positive or negative charge. Solar ultraviolet radiation is the principal reason neutrals become ionized in comets.

METEOR: The rapidly moving streak of light caused by a particle as it burns up in the earth's atmosphere.

METEORITE: A natural particle reaching the surface of the earth from space after traveling through the earth's atmosphere.

METEOROID: A natural particle in space before it enters the earth's atmosphere.

NODE: Point of intersection; see Ascending node and Descending node.

NUCLEUS: The inner part of the comet's head, the source of all cometary phenomena, the nucleus is believed to be a snow ball of frozen gases and dust.

OPPOSITION: The position of the comet situated in a plane
containing the sun and earth at celestial longitude 180
degrees from the sun and crossing the meridian at midnight.
When the comet is at opposition it is nearest the earth and
in a favorable position for observation.

PERIHELION: The point in the comet's orbit around the sun
which is closest to the sun.

PERTURBATION: Gravitational effects upon the   orbital
motion of the comet by masses other than the sun (usually
other planets).

PLASMA: A gas of positive and negative ions.

RETROGRADE ORBIT/ROTATION: The comet's orbit and rotation
are in a clockwise direction rather than the more usual
counter-clockwise (as viewed from the north side of the
solar system).

TAIL: The general term used to describe the ejecta (ions
and dust) streaming out from the comet head opposite the
sun.

ZODIACAL LIGHT: A general glow throughout the sky caused
by sunlight scattered by interplanetary dust.   It is
brightest near the sun.

Author Index

Franklin, Kenneth L., 1115
Freitag, Ruth S., 1263, 1279
Friedlander, A.L., 1024,
  1025
Friedman, L., 1191
Frost, Edwin B., 877, 955,
  975
G., J., 8, 46
Gamma Draconis, 74
Gemsege, Paul, 4
Gerard, J., 171
Giacobin, Michael, 570
Gill, David, 571
Glackens, L.W., 632
Glasgow, Thurman A., 1021
Glasmachers, A., 1213
Goatcher, A. Winton, 465
Gore, J.G., 281
Graham, George Cornelius,
  115
Green, Daniel W.E., 1232
Griffin, Frank Loxely, 466
Griffith, W. Branford, 870
Griggs, H.W., 381
Grover, C., 166
Haimson, Leonie, 1133
Hajduk, A., 1288
Hall, Maxwell, 863, 983
Halliday, Ian, 1272
Hallock, William, 282
Hanner, Martha S., 1142
Hast, S.L., 1065
Hawks, Ellison, 467
Hechler, F.W., 1183
Hechler, M., 1183
Hedrick, John T., 905
Hendrie, Michael J., 1231
Henkel, F.W., 173, 176,
  179
Herschel, John F.W., 118,
  126, 130, 1018
Heward, Ed. Vincent, 229
Hicks, E. Rupert, 306
Hind, J. Russell, 135
Hirayama, K., 368
Holetschek, J., 199
Horner, G.R., 945
Horsewood, J.L., 1067
Houpis, Harry L.F., 1127
Hughes, C.E., 351
Hughes, David W., 1056, 1068,
  1085, 1087, 1103, 1262,
  1282
Humphreys, W.J., 538, 929,
  976
Hunter, A.F., 468

I., B., 7
Innes, D.M., 442
Innes, R.T.A., 921, 939
Irwin, Wallace, 488
Ivanov, K.G., 1010
Izzard, W.H., 155
Jacoby, Harold, 268, 539
Jacobson, Robert A., 1071,
  1072
Jenkins, J., 864
Jones, D.R.L., 1019
Joslin, B.F., 124, 125
Kaempffert, Waldemar, 252,
  402, 418, 793
Kane, Van R., 1175, 1246
Keeling, B.F.E., 947
Kiang, Tao, 1027, 1032, 1200
Kitfield, James, 1296
Kron, E., 956
Klein, Jerry, 1006
Kronk, Gary, 1187
Lampland, C.O., 308
Large, Arlen J., 1088, 1186
Larkin, Ralph B., 225
Larson, S.M., 1298
Laufer, Joseph, 1217, 1290
Leach, C., 489
Ledger, E., 168, 846
Lee, Oliver J., 215, 244
Lee, William Ross, 935
Lenorovitz, Jeffrey, 1155,
  1197
Leon, Luis G., 891
Leuscher, A.O., 898
Leverin, Albert, 423
Littrow, C.L., 83
Loomis, Elias, 121
Lowell, Percival, 360, 906,
  948
Lubbock, J.W., 24, 25, 49
Lynch, James K., 452
Lynn, W.T., 142, 143, 145,
  149, 161, 164, 169, 177,
  190, 222, 380
M., N., 6
McAdam, D.J., 540
McCready, Kelvin, 276
MacDonnell, W.J., 318
McGill, H.H., 859
McGillivray, D., 915
McIntosh, B.A., 1288
McLaughlin, William I., 1176,
  1184
Maclear, Thomas, 123, 127,
  129
McNiece, Mili Ve., 1188

McPike, Eugene Fairfield, 154, 158, 159, 160
Mammano, Augusto, 1029
Mann, F.I., 1067
Maran, Stephen P., 1198
Margrave, Thomas E., Jr., 1046
Marsden, Brian G., 1041, 1076
Matkiewitsch, L., 224
Mayer, Hy, 329, 420, 501, 781; see Illustrations 4, 5, and 6
Me, 22
Meadows, A.J., 1019
Mendis, D.A., 1127
Mendis, G.D., 1082
Merfield, C.J., 319, 396, 916
Metcalf, Joel H., 226, 892
Meyer, W.F., 299
Michielsen, H.F., 1017
Millard, Bailey, 878
Mitchell, S. Alfred, 320, 339, 397, 469, 572, 847
Moir, James, 880
Moorehouse, D.W., 893
Morris, Charles E., 1232
Morris, Charles S., 1291
Morrison, Richard James, 108
Mueller, M.J., 132
Mullens, E.T., 717
Mumford, George S., 1011
Munckley, Nicolas, 11
Murray, Bruce, 1246
Mutch, Thomas A., 1108, 1109
Nankive, Frank, 513
Naulty, Edwin Fairfax, 541
Nauticus, 70
Neill, G.A., 865
Neward, E.V., 148
Newburn, Ray L., 1143, 1144, 1191, 1214
Ney, Edward P., 1220
Norman, Colin, 1059
Nye, Bill, 447
O'Halloran, Rose, 917
Olbers, Heinrich W.M., 28, 37
Olivier, Charles P., 340, 961
Olson, Roberta J.M., 1083
O'Neill, H.C., 321
Oppenheimer, Michael, 1133
O'Toole, Thomas, 1168, 1268, 1289
Overbye, Dennis, 1201
Payn, Howard, 869
Pearson, D., 98
Pearson, John F., 1047

Perrine, C.D., 907, 922, 925, 950
Petri, E., 881, 901
Phillips, Theodore E.R., 262, 277, 286, 369, 470, 848
Philomathes, 23
Pickering, Edward C., 382
Pickering, William H., 370, 398
Plana, 63
Plummer, H.C., 997
Pontecoulant, Louis G., 34
Porter, Elizabeth Crance, 821
Proctor, Mary, 165, 449, 497, 502, 506, 514, 556, 573, 594, 633, 747, 771, 782, 825, 830, 832, 835, 985, 995
Pournelle, Jerry, 1202
R., M.H., 807
Rahe, Jurgen, 1193
Rambaut, A.A., 406, 928
Raulein, Theodore M., 245
Raven-Hill, L. 812
Ravene, Gustave, 147
Rayl, G.J., 1057
Raymond, W.E., 930
Redding, Cyrus, 62
Redfearn, Judy, 1163
Reeve, Arthur B., 322
Reeves, Paschal, 1016
Rehn, Frederick J., 542
Reinhard, R., 1104, 1143, 1145, 1194, 1215
Rexroth, Kenneth, 1002
Rich, Vera, 1173
Richards, L. Adolph, 209
Richardson, Robert S., 1012, 1015
Rigge, William F., 202, 371, 399
Roberts, Alexander W., 227, 263
Roberts, Ruel W., 331
Ronan, C.A., 1287
Rosenberger, Otto August, 90
Roosen, Robert G., 1041
Ross, David, 940
Rudaux, Lucien, 574
Russell, Henry Norris, 419, 498, 543
Rutherford, F. James, 1249A
Rutledge, Archilbald, 343
Sagan, Carl, 1112
Salisbury, David F., 1203
Salmon, W.H., 996

Saunders, T. Bailey, 854
Schaeberle, J.M., 575
Schove, D. Justin, 1000, 1001
Schwarzschild, K., 956
Scomp, Henry Anselm, 280
Scott, Samuel, 1004
Seagrave, F.E., 894, 965
Seaman, Owen, 237
Seares, John, 35
Searle, George M., 197, 278
See, T.J.J., 400, 471
Sekanina, Zdenek, 1013, 1086,
    1146, 1199, 1275, 1298
Serviss, Garrett P., 230
Shatraw, Milton, 1020
Shaw, H. Knox, 951
Shevnin, A.D., 1010
Singer, S. Fred, 1189
Slipher, V.M., 308, 933
Sloane, James, 77
Slocum, Frederick, 949
Smart, David, 150, 162, 163,
    212, 957, 967
Smith, Lucia E., 401
Smith, Sherwin D., 1036
Smyth, W.H., 109, 120
South, James,, 48, 52, 53,
    56, 65, 67, 73, 75, 80, 84,
    85, 86, 93, 97, 103
Sperra, William E., 895
Stanton, R.H., 1147
Stebbins, Joel, 900
Stein, J., 849
Stenquist, D., 881, 901
Stephenson, Bill, 999
Stone, Greg, 1038
Stratford, W.S., 73, 100,
    102, 104
Strout, Richard L., 1264
Sullivan, Walter, 1130
Swartz, Helen M., 896
T., 99
Tatum, Jeremy B., 1170, 1235,
    1285
Taylor, C.S., 352
Taylor, John, 101, 107
Taylor, T.G., 128, 131
Tebbutt, John, 323, 936
Tennyson, Alfred, 140
Terry, Sara, 1174
Tessier, A.C., 448
Thompson, Arthur, 309
Thompson, C., 57
Thornton, Catherine L., 1071,
    1072
Tittman, O.H., 902

Todd, David, 310, 919
Touchstone, 234
Toynbee, Paget, 966
Trachtenberg, Jeffrey A.,
    1297
Trendelenberg, E.A., 1216
Trogus, Wolfgang, 1148
Tsu, Wen Shion, 994
Turner, Herbert H., 184, 794,
    822, 962
Twain, Mark, 182
Valentry, Duane, 1227
Vanysek, V., 1013
Very, Frank W., 979
Veverka, Joseph, 1110
Visconti, G., 1031
Waldrop, M. Mitchell, 1196,
    1230
Wallis, Max K., 1077
Warner, Irene E. Toye, 279,
    941
Washburn, Mark, 1153
Watson, A.D., 213
Watson, Fletcher G., 1051
Wayman, P.A., 1032
Weeks, Albert L., 1039
Wekhof, A., 1149
Wells, W.C., 1025
Wendell, O.C., 186
Wesley, John, 1
West, J.L., 1055, 1064
Wheeler, Robert L., 998
Whipple, Fred L., 1150
Whitmell, C.T., 341
Whitney, Mary W., 908
Wilczewski, Joseph, 472
Wilford, John Noble, 1049,
    1099, 1180, 1251, 1253,
    1299
Will, George, 1185
Williams, Gaar, 634, 748
Williams, Gurney, III, 1121
Williams, John, 138
Willis, Edgar C., 264
Wilson, H.C., 193, 200, 403,
    909, 918
Wilson, Ralph E., 850, 909
Winter, Patty, 1223
Winters, William H., 635
Winthrop, John, 12, 17, 21
Wisterman, 636, 718
Wood, H. John, 1195
Wood, Lincoln J., 1065, 1177,
    1229
Wooden, William H., II,
    1048, 1105

Wright, W.H., 210
Wurm, Karl, 1029
Yeomans, Donald K., 1045,
  1052, 1152, 1200, 1214,
  1274, 1275

Yoke, Ho Peng, 1007
Zornlin, Rosina Marina, 36
Zwack, George M., 304

Subject Index

Accra, 836, 870
Acetylene gas, 266
Aden, 738
"Adoration of the Magi"
  (Giotto), 1083
Advertisements, 355, 404,
  412, 450, 456, 525, 551,
  767, 999, 1037, 1279;
  see Illustration 10
Aitken, John, 913
Alabama, Lawrence County,
  629; Mobile, 27, 500, 523,
  605; Montgomery, 560, 792;
  Talladega, 789
Alexander the Great, 107
Alfred lecture, 485, 822
Algiers, 923, 935; Bouzareh,
  457
Alighieri, Dante, 969;
  "Convivio," 966
Altitude (of comet), 1170
American Institute of
  Physics, 1228
Aneroid barometer, 644
Anglo-Saxon Chronicle, 593
Annular micrometer, 109, 120
Aphelion, 1161
Apparitions, specific (see
  History for general
  discussions of the various
  apparitions), 141 A.D.,
  147; 760 A.D., 194; 837
  A.D., 194; 1066 A.D., 54,
  545, 593, 849; 1222 A.D.,

62, 185, 187; 1531 A.D.,
  854; 1682 A.D., 1022
Aquarid meteor showers, 151,
  152, 172, 192, 961, 1052,
  1288, 1301
Aquarius, 23
Arago, Francois, 30, 92, 110
Argentina; Buenos Aires, 117;
  Cordoba, 907, 910, 922,
  925, 950
Arianespace, 1234
Aristotle, 305
Arizona; Flagstaff, see
  Lowell Observatory; Tucson,
  474, 620, 678
Arizona, University of, 474,
  813
Ark of the Covenant, 280
Art, 1004, 1083, 1092
Ascending node, 243, 272
Astronomical and Astrophys-
  ical Society of America,
  259, 330, 912, 914, 971,
  972, 973, 974, 976
Astronomische Gesellschaft;
  Lindemann Prize, 156
Atmospheric conductivity, 538
Aurora, 538, 562, 569, 612,
  643, 659, 689, 737
Australia, 318, 940;
  Melbourne, 271, 836; Perth,
  415; Sydney, 930; Windsor,
  936
Austria, 1236; Vienna, 415

1269
Chile, 441; Santiago, 457, 883
China, 135, 136, 138, 147, 185, 425, 435, 872, 915, 938, 994, 1007, 1027, 1087
Chinese records, 135, 136, 138, 147, 185, 994, 1007, 1027, 1081, 1087
Christian Literature Society, 915, 938
Christie, W.H.M., 858
Clarkson Memorial School of Technology (NY), 294
Clemens, Samuel L., 434, 1000, 1005
Climate, effects on, 38, 122, 520, 538, 546, 606, 667, 721, 856, 878, 903, 929
Clubs, comet, 661, 734
Cocktails (named for), 1006, 1037
Collision (earth and comet), 1, 2, 4, 42, 94, 195, 203, 230, 252, 274, 294, 303, 304, 315, 372, 377, 384, 398, 418, 435, 487, 519, 539, 560, 584, 678, 1009, 1139, 1171
Colorado; Aspen, 578; Denver, 654, 664, 686, 723, 735, 762; Greeley, 724; Pueblo, 668
Coma, 250, 840, 844, 932, 957, 967, 1069, 1077, 1085, 1141, 1195, 1245; Ionization velocity, 1212
Comet Crommelin, 1299
Comet parties, 481, 482, 493, 494, 536, 581, 661, 680, 696, 704, 710
Committee on Comets, American Astronomical and Astrophysical Society, 204, 259, 291, 313
Comparison orbits, 104
Comparison stars, see Stars, comparison
Composition (comet), 547
Conic sections, 295
Connecticut; New Haven, 59, 66, 121, 222, 619
Constantinople, 30, 110, 142, 144, 149, 171, 198, 202, 216, 232, 249, 334, 399, 732, 811, 814

Coronas, 976
Cowell, P.H., 219, 254, 278, 353, 786, 943
Cox, Simon H., 749, see also 631
Crommelin, comet, 1299
Crommelin, A.C.D., 219, 254, 278, 311, 330, 353, 387, 448, 505, 509, 786, 839, 920, 943, 962
Cuba; Havana, 328
Curacao, 422; Willemstad, 738
Curtis, Heber D., 991
Curtiss, F. Homer, 986, 987
Cyanogen (gas), 244, 345, 346, 508, 510, 572, 574, 912, 1283
Czechoslovakia, 1236
Damoiseau, Marie Charles, 57
Danielson, G. Edward, 1250, 1251, 1253, 1254, 1257, 1258, 1262, 1265, 1267, 1270
Dante Alighieri, 969; "Convivio," 966
David (King of Israel), 280, 1001
Declination, 50, 91, 130, 131, 287
Delambre, Jean-Baptiste, 157
Delaware; Wilmington, 91, 95, 496
Descending node, 7, 21, 272, 323
Diameter (comet), 83, 241, 348, 517, 524, 898, 1085
Dines, W.H., 903
el Din, Hassan Fahmy Gamal, 960
Dinwiddie, A.B., 504
Discoveries, first, see Recovery
Disasters (attributed to), 38, 108, 110, 164, 274, 279, 418, 986, 987, 1137, 1139, 1227
Disturbance action, 671
Dominion Observatory, 453
Doolittle, Charles L., 743, 772
Dornier-Werke, G.M.B.H., 1116
Drake University, 609
Drama, 749
Drinks (named for), 1006,

1037
Dublin Observatory, 57
Dust (cometary), 292, 627,
  1029, 1031, 1069, 1082,
  1086, 1104, 1140, 1141,
  1143, 1144, 1146, 1247,
  1282, 1292
Dynamical processes, 1127
Earth currents, 538
Earthquakes, 38
Eccentricity, 214, 243
Eclipses, 119; Lunar, 287,
  588, 664, 783, 795, 800,
  805, 806, 817, 954; Solar,
  1018
Edinburgh (Scotland), Royal
  Society of, 913
Edward, VII, King of England,
  503
Egypt, 886; Cairo, 960;
  Helwan, 794, 816, 850, 923,
  927, 947, 951, 958
Elections, British Parlia-
  mentary, 335, 485
Electrical particles, 516,
  538, 568, 768, 819, 881
Electrical Waves, 538, 768,
  819, 881
Elecrometer, quadrant, 309
Elements, 24, 25, 35, 193,
  242, 243, 295, 1019, 1152,
  1275
Elliptical coordinates, 342
Ellerman, Ferdinand, 259,
  330, 912, 914, 971, 972,
  1003
von Encke, Johann Franz, 105
Energetic particles, 1077
Energy (of comet), 1205
England, 15, 140, 220, 232,
  314, 548, 715, 796, 837;
  Abingdon, 312; Bath, 312;
  Bristol, 60, 845; Bedford,
  109; Cambridge, 61, 133,
  362, 448, 732, 784, 912;
  Ealing, 312; Grahamstown,
  317, 338, 394, 464;
  Greenwich, 441, 732, 946;
  Hampstead, 784, 796;
  London, 73, 85, 87, 461,
  679; Manchester, 1284;
  Norwich, 264; Oxford, 406,
  928; Rye, 448
Envelope, 671
Ephemerides (1835), 26, 37,
  46, 67, 84, 85, 90, 97,

99, 103, 104, 107; (1910),
  150, 186, 199, 207, 214,
  218, 224, 240, 241, 256,
  269, 319, 320, 332, 333,
  335, 340, 347, 363, 366,
  383, 391, 425, 837, 924,
  953, 965; (1986), 1023,
  1102, 1152, 1271, 1275
Esclagnon, M.E., 980
European Space Agency, 1070,
  1080, 1085, 1090, 1095,
  1104, 1116, 1118, 1119,
  1122, 1123, 1124, 1125,
  1126, 1128, 1130, 1135,
  1145, 1148, 1151, 1154,
  1155, 1157, 1160, 1165,
  1193, 1194, 1207, 1208,
  1213, 1215, 1221, 1234,
  1245, 1253, 1256, 1259,
  1281, 1283, 1292, 1293
Evershed, John, 838, 978
Excommunication (of comet),
  see Callixtus III, Pope
Exorcism (of comet), see
  Callixtus III, Pope
Explorer 3 spacecraft, 1294
Expressionism, German, 1063
Fiction, 182, 1138
Finmark, 790
Fitzgerald, John F., 585
Flamsteed, John, 148, 180,
  190
Florida; Tampa, 535, 584,
  667, 703
Fischer, Johann Karl, 32
Flyby technique, 1014, 1094,
  1122, 1145, 1177, 1233
Flythrough technique, 1024,
  1065, 1070, 1077, 1086
Forbes, A.H., 521
Foreigners (in U.S.), 432,
  672, 682, 686; Chinese,
  652; Italians, 648, 655,
  672
Fowler, A., 912
France, 520, 618, 1131,
  1132, 1164, 1173, 1223,
  1236; Boulogne-sur-Seine,
  851; Marseilles, 851, 874;
  Meudon, 271; Nice, 923;
  Paris, 15, 92, 422, 546,
  570, 579, 619, 692, 704,
  732, 738, 777, 855; Pic du
  Midi d' Osseau, 884
French Space Agency, 1131,
  1132, 1164, 1173, 1223,

271

1099, 1107, 1108, 1109,
1110, 1111, 1112, 1134,
1135, 1174, 1175, 1179,
1180, 1182, 1196; Private,
1162, 1226; see also
Probes, space
Missouri; St. Louis, 566,
619, 644
Mitchell, S. Alfred, 533
Mithradates, 107
Mizar, 82
Models (of comet), 306, 426,
452, 544, 1051, 1217
Moebius, August Ferdinand, 32
Moon, 287, 588, 589, 783,
795, 800, 805, 806, 817,
954
Moore, Edmund B., 628
Morehouse, D.W., 609
Morrison, Richard James, 164
Motion (of comet), 298, 1023,
1026, 1052; history, 1214;
irregularities, 1013;
long-term, 1200; mean, 916
Mt. Hamilton, 340
Mt. Palomar Observatory,
1250, 1251, 1253, 1254,
1255, 1257, 1258, 1262,
1265, 1267
Mt. Wilson Observatory, 547,
626, 737, 766, 773, 882
Mt. Whitney, 954
Mueller, M.J., 113
Murray, Bruce, 1246
Mutch, Thomas A., 1111
NASA; see National Aero-
nautics and Space Admin-
istration
National Academy of Sciences
(U.S.), 914
National Aeronautics and
Space Administration, 1049,
1053, 1059, 1073, 1078,
1080, 1088, 1090, 1091,
1094, 1098, 1099, 1108,
1109, 1110, 1111, 1112,
1113, 1125, 1126, 1128,
1134, 1135, 1151, 1157,
1159, 1174, 1179, 1180,
1184, 1203, 1216, 1225,
1278
National Science Foundation,
1228
Naulty, Edwin F., 557, 565
Naval Observatory, see U.S.
Naval Observatory

Nebraska; Davis City, 818
Negative ions, 1149
Negroes, 503, 523, 605, 611,
622, 632, 645, 663, 681,
682, 699, 700, 735
Neptune, 968, 1270; see also
Perturbations
Nevada; Carson City, 429,
726, 761
New England, 1038
New Haven Herald, 59
New Jersey; Carlton Hill,
731; Mt. Holly, 532;
Newark, 709; Roselle, 670;
Towaco, 451; Woodbury, 499
New Mexico; Albuquerque, 653
New York; Brooklyn, 481;
Geneva, 287, 511, 766;
Hyde Park, 712; New York
City, 175, 438, 449, 475,
491, 492, 497, 514, 528,
536, 554, 672, 734, 745,
747, 771, 776, 798, 813,
826, 828, 830, 832, 985;
Poughkeepsie, 908;
Schenectedy, 88, 124, 125
New York Times, 744, 756,
1006
New Zealand, 945
Newburn, Ray L., 1244
Newfoundland, 999; St.
John's, 422
Newsletters, 1217, 1218
Newton, Isaac, 12, 17, 168,
184, 389, 850, 997
Nihongi, 368
Nihonsyoki, 368
Nippoldt, Dr., 899
Nongravitational forces,
1045, 1143
Norman invasion (1066 A.D.),
140, 220, 232, 314, 548;
see also Bayeux tapestry
Norris, Ray, 1284
North Carolina; Ashville,
699
North Dakota; Fargo, 680;
Mandan, 697; Minot, 817;
Grand Forks, 733
Norway, 427
Novels, 1138
Nucleus, 10, 80, 83, 86,
250, 271, 290, 300, 454,
483, 524, 742, 827, 840,
844, 884, 932, 944, 957,
960, 964, 967, 1013, 1043,

274